5G and Beyond Wireless Communication Networks

5G and Beyond Wireless Communication Networks

Haijian Sun
University of Georgia
Athens, GA, USA

Rose Qingyang Hu
Utah State University
Logan, UT, USA

Yi Qian
University of Nebraska-Lincoln
Omaha, NE, USA

Registered Offices

John Wiley & Sons, Inc., 111 River Street, Hoboken, NJ 07030, USA

John Wiley & Sons Ltd, The Atrium, Southern Gate, Chichester, West Sussex, PO19 8SQ, UK

For details of our global editorial offices, customer services, and more information about Wiley products visit us at www.wiley.com.

Library of Congress Cataloging-in-Publication Data applied for:

Hardback ISBN: 9781119089452

Cover Design: Wiley
Cover Image: © timandtim/Getty Images

Set in 9.5/12.5pt STIXTwoText by Straive, Chennai, India

Contents

About the Authors

Haijian Sun, PhD, is an Assistant Professor in the School of Electrical and Computer Engineering at the University of Georgia, Athens, GA, USA.

Rose Qingyang Hu, PhD, is an Associate Dean for Research in the College of Engineering and a Professor in the Department of Electrical and Computer Engineering at Utah State University, Logan, UT, USA. She is a Fellow of IEEE.

Yi Qian, PhD, is an IEEE Fellow and is a Professor in the Department of Electrical and Computer Engineering at the University of Nebraska-Lincoln, Lincoln, NE, USA. He is a Fellow of IEEE.

Preface

Motivated by the increasing demands for connectivity, modern wireless technologies have experienced rapid developments. Efforts from academia, industry, and government have pushed wireless research at an unprecedented level. As a complex solution, wireless systems comprise many components, from physical layer, to network and upper application layer. One of the most exciting innovations in the past decade is 5G physical layer, which includes new radio (NR), new spectrum, coding, etc. Inspired by recent physical layer research advances in 5G and beyond wireless systems, this book intends to present the state-of-art challenges and solutions for physical layer techniques that are already applied, or will be utilized in wireless systems. This book covers a variety of topics, primarily on the intersection of 5G and beyond system with NR, mobile edge computing, and machine learning, and spectrum sharing. Ultimately, we expect to deliver a more energy-, spectral-, and computation-efficient wireless technology.

There are twelve chapters in this book. They can be categorized into three main topics. Chapters 1–6 focus on 5G new radio research, especially recent advancements and systematic research on non-orthogonal multiple access (NOMA). Chapters 7–9 discuss the interactions of mobile edge computing and wireless technology. Chapters 10 and 11 focus on secure spectrum sharing in 5G and beyond era. Chapter 12 concludes this book and further discusses some future research directions. Below, we briefly summarize each chapter.

Chapter 1 presents an overview of 5G and beyond wireless system. We start by introducing system requirements and their technical challenges. Then the enabling technologies from NR, mobile edge computing, and heterogeneous communication architecture are illustrated.

Chapter 2 discusses the integration of 5G networks with device-to-device (D2D) communication. Specifically, the 5G system with underlaid D2D is presented. We show that such system can increase spectral efficiency, providing that resource allocation is properly designed.

Chapter 3 deals with NOMA-enabled practical wireless networks. The highlight is the integration of error propagation, a well-known issue in NOMA. It shows that error propagation can degrade system performance, depending on the residual value.

Chapter 4 presents 5G relay and IoT networks with NOMA. In the first part, we derive the outage probability in the relay system and show the potential of such a configuration. Then, in the second part, the IoT network with power transfer capability is considered.

Chapter 5 discusses the robust beamforming problem in cognitive radio system; we specifically illustrate the beamforming design when bounded channel estimation error is present.

Chapter 6 is a continuation of Chapter 5. It considers a more realistic channel estimation model, in which channel estimation error is modeled as Gaussian variable. Correspondingly, beamforming design also changes.

Chapter 7 presents mobile edge computing in 5G wireless networks. The system aims at reducing computing latency and offload computation tasks to nearby edge servers. With the goal of maximize computation efficiency, resource allocation optimization is proposed and designed.

Chapter 8 further considers security enhancements in mobile edge computing. Our security design focuses on physical layer, i.e. from wire-tap channel perspective.

Chapter 9 deals with an innovative wireless system to facilitate distributed machine learning as opposed to machine learning for wireless communication. We show that efficient information exchange via wireless can accelerate large-scale distributed machine learning. A direct application is wireless federated learning.

Chapter 10 provides an overview for secure spectrum sharing with machine learning. While secure spectrum sharing is not a new topic, we have witnessed advancements in this area, especially with machine learning techniques.

Chapter 11 presents detailed machine learning methodologies for secure machine learning. This chapter illustrates several dominant attacks and their respective mitigation approaches.

Chapter 12 concludes this book and gives some emerging research directions in 5G and beyond wireless networks.

We hope our readers will enjoy this book.

January 2023

Haijian Sun, University of Georgia
Rose Qingyang Hu, Utah State University
Yi Qian, University of Nebraska-Lincoln

Acknowledgments

We would like to thank our families for their continuous support and love.

We would like to express our sincere gratitude and appreciation to our colleagues, students, and staff at University of Georgia, Utah State University, and University of Nebraska-Lincoln who have supported us throughout the journey of writing and publishing this book. Your encouragement, feedback, and advice have been invaluable in shaping the final product, and we are truly grateful for your contributions.

We also would like to extend our heartfelt gratitude to Juliet Booker, Sandra Grayson, and Becky Cowan at Wiley who have played an integral role in bringing our new book to fruition. Thank you for your hard work, support, and guidance throughout the publishing process.

This book project was partially supported by the U.S. National Science Foundation under grants CNS-2236449, CNS-2007995, CNS-2008145, ECCS-2139508, and ECCS-2139520.

Haijian Sun, Rose Qingyang Hu, and Yi Qian

1

Introduction to 5G and Beyond Network

We have witnessed an unprecedented development of wireless technology for the past few decades. Starting from 1980s, when the first mobile phone was released, major wireless technology advanced almost every decade. From first generation (1G) to 4G. The invention of smart devices, such as phones, tablets, and home appliances, is the main driving force for the ever-increasing mobile traffic today. It is not surprising that mobile traffic increased 10-fold between 2014 and 2019 globally. The mobile data traffic is expected to grow much faster than fixed IP traffic in the upcoming years [34]. Wireless technologies dramatically changed the way people interact, communicate, and collaborate, especially at post-Covid era. The need for faster, more efficient and secure, and intelligent communication technique remains strong. While the current wireless communication systems such as 4G long term evolution (LTE) have been pushed to their theoretic capacity limit, different air interface and radio access technologies including heterogeneous network (HetNet) [76, 77], multiuser multi-input multi-output (MU-MIMO) [105], and device-to-device (D2D) communication [51] have become potential paradigms to fulfill the gap between demands from end users and the capacity that current air interface can provide.

1.1 5G and Beyond System Requirements

In their pioneering work [10], Andrews *et al.* evaluated the requirements for 5G. In short, 5G wireless communication system should provide 1,000 times aggregate data improvement over 4G, support for as low as 1 ms round-trip latencies, 10 times longer battery life for low-power devices, and also support 10,000 times or more low-rate devices in a single macro cell, see Figure 1.1 for a brief illustration. Due to those high requirements, the transformation from 4G to 5G cannot be simply fulfilled by extensions of current technologies. In general, 5G and beyond system should support or deliver the following aspects. Notably, (i) more bandwidth. Currently commercial cellular systems use frequencies below 6 GHz (sub-6 GHz); in fact, there is abundant bandwidth in the millimeter-wave (mmWave) band, for example in 28 GHz and above, which can provide more bandwidth that previously have not been applied in cellular networks. (ii) More antennas. Higher frequency also brings smaller form factor of large antenna arrays. Additionally, the signal processing techniques in terms of massive MIMO and transceiver design also improved

5G and Beyond Wireless Communication Networks, First Edition. Haijian Sun, Rose Qingyang Hu, and Yi Qian.
© 2024 John Wiley & Sons Ltd. Published 2024 by John Wiley & Sons Ltd.

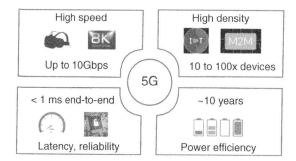

Figure 1.1 Four main goals for 5G.

significantly. (iii) New radios (NR). The physical layer in 5G will change dramatically, specifically the 5G NR, which includes the new multiple access technology, the new air interface, and a combination of several existing techniques. (iv) New schemes. It is expected that ultra dense networks (UDN) will be heavily deployed. The density of small base station (BS), such as micro BS, femto cell, and pico cells, will be much higher than that in 4G. But they share the similarity in terms of deploying BSs with different powers to provide seamless coverage, as well as performance improvements from short-range communications. (v) High intelligence. It is expected that beyond 5G systems should support higher level of intelligence. Emerging applications such as Artificial intelligence (AI), semantic communication, and robots will surely benefit from AI-friendly wireless technology. (vi) Pervasive wireless. It is anticipated that each person will carry more personal devices for enhanced life style and health monitoring. To support ubiquitous wireless connectivity, those devices need be connected. Current network architecture can hardly support such high number of devices simultaneously.

1.1.1 Technical Challenges

The above promising technologies are able to deliver ambitious goals of 5G, but they ultimately encounter some challenges. First of all, even though high-frequency bands have major vacancy, mmWave signals are notorious for weak penetration and vulnerable blockage; hence, the transmission characteristics are big concerns. Moreover, studies also have shown mmWave signals have high attenuation due to atmospheric gaseous, rain, concrete structure, glasses, even foliage. The real-world deployment of such mmWave systems needs to be carefully studied and planned. Secondly, from the transceiver design perspective, higher-frequency signals impose challenges in circuit design, materials, and heating issues. Nyquist theorem sets the lower boundary for sampling rate in communication systems. With wide bandwidth in mmWave spectrum, sampling rate can reach up to 10 Gbit/s level, and high-speed circuit design becomes very difficult. It is also reported that the energy efficiency for components (power amplifier, analog-to-digital converter, digital-to-analog converter) in high frequency is low, only around 10%. One of the major concerns from network operators is that power consumption will hike due to 5G. Furthermore, the low efficiency in these components also brings thermal issues in hand-held devices, degrading user experiences. Thirdly, with mmWave band, performance gain largely comes from large-scale antenna array, current design can integrate hundreds of antenna elements in a small area (due to small wavelength of mmWave signals). Even though this can facilitate the beamforming, which generates narrow but stronger signals toward desired direction, the overhead for channel estimation, precoding, and beam tracking is too large. Fourthly, in UDN networks, since the transmitter density is high, signals can cause higher interferences with each other. The problem will be more severe with high-density users in the same area. Challenges in mobility management, interference management, and heterogeneity nature of devices are severe. Lastly, it is expected to support intelligent applications in beyond 5G systems. For example, conventional communication systems are transparent of message (i.e. they are only responsible for transmitting bits but do not know any further info). Semantic communication, on the other hand, has knowledge of the underlying message, and the communication scheme can be dynamically changed to fit different needs of the message. Besides, ubiquitous wireless signals open door for sensing applications, such as localization, monitoring, and healthcare. In recent years, intelligent communication system has been proposed to accommodate these needs. A notable example is wireless federated learning system to cater the distributed machine learning. However, a deep integration from wireless design perspective is strongly desired.

Recently, there are several emerging technologies which aim to deliver the goal of 5G and beyond, and address the challenges above. Specifically, in this book, our focus is on the physical layer techniques, such as 5G NR non-orthogonal multiple access (NOMA) and physical layer (PHY) mobile edge computing (MEC), high-level communication architecture for pervasive Internet of Things (IoT) devices, as well as wireless federated learning system design. We have conducted preliminary researches to address the challenges mentioned above. Specifically, we discuss how to utilize NOMA on improving aggregated data rate and supporting more devices simultaneously, propose schemes for wearable IoT communications, discuss the usage of MEC on helping with power consumption and latency, and analyze how wireless design can facilitate distributed machine learning. Below we briefly introduce each enabling technique.

1.2 Enabling Technologies

1.2.1 5G New Radio

1.2.1.1 Non-orthogonal Multiple Access (NOMA)

Initially proposed by NTT DOCOMO as an enhancement for LTE-advanced (LTE-A) in 2013, NOMA has been recognized as one of the most promising techniques for 5G due to its capability of supporting a higher spectral efficiency (SE) and native integration of massive connectivity. The basic principle of NOMA is that at the transmitter side, multiple signals are added up with different powers, forming a superimposed signal (SS). To ensure weak user's quality of service (QoS), at the receiver side, successive interference cancellation (SIC) is used to retrieve each user's signal sequentially from the SS. Specifically, a user can decode the strongest signal by treating other signals as interference. If the decoded signal is its own data, SIC stops. Otherwise, the receiver subtracts the decoded signal from the SS and continues to decode the next strongest signal. Notice that SS with SIC is not new; in information theory, this duo is a capacity-achieving technique in the uplink communication. However, the difference is in NOMA, the weak user has a stronger power, which

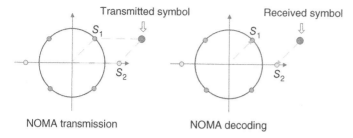

Figure 1.2 NOMA principles: transmission and decoding stage.

is not information-theoretic optimal. Since its design philosophy may be combined with diverse transceivers, it has drawn tremendous attention in multiple-antenna systems and in downlink and uplink multi-cell networks. In contrast to classic orthogonal multiple access (OMA), NOMA provides simultaneous access to multiple users at the same time and on the same frequency band, for example by using power-domain multiplexing. It has been shown that NOMA is capable of achieving a higher SE and energy efficiency (EE) than OMA, such as OFDMA, time division multiple access (TDMA), and frequency domain multiple access (FDMA). Figure 1.2 shows the basic principle of NOMA with data encoding and decoding. S_1 and S_2 are the symbols for users 1 and 2, respectively. We also assume user 1 has a better channel than user 2. At the transmitter side, binary phase shift keying (BPSK) and quadratic phase shift keying (QPSK) modulation are applied, respectively, for the two users. Clearly, the average symbol power of S_2 is larger to compensate for the unfavorable channel. Actual transmitted symbol is simply the addition of these two. At the receiver side, symbols with the highest power will be decoded first, in this example, S_2. Besides, since the received symbol is on the right side of y-axis, for BPSK, it will be decoded as S_2, and then removed from the composite signal, which only has S_1 left. Notice that the symbols can use the same modulation scheme as long as they have different power. Most NOMA works, however, do not consider any specific modulation, rather they apply the Gaussian coding and perform analysis based on information-theoretic perspective.

The disadvantage of NOMA, however, lies in the following aspects. Firstly, NOMA requires a more complicated receiver structure to perform SIC; hence, the cost will be higher and receiver architecture will also be changed accordingly. Secondly, during SIC procedure, one user will decode signal from others; this will cause security and privacy concerns. Lastly, depending on implementation, this successive decoding will impose certain delays for users.

Starting from 3rd Generation Partnership Project (3GPP) LTE Release-13, NOMA, as one of the techniques in multi-user superposed transmission (MUST), has become part of the standardization. In 2017, with LTE Release-14, 15 MUST schemes have been proposed for the uplink NR. Additionally, NOMA has attracted extensive attention from industry. NTT DoCoMo and MediaTek collaborated to have a field test of NOMA in Nov. 2017. With a simple scenario of one base station and three users, they were able to achieve 2.3 time spectral efficiency compared with current technology.[1]

---

1 MediaTek Newsroom, Nov. 2017.

Nevertheless, we have applied NOMA in many schemes and systematically studied its performance, for example NOMA with D2D, with MIMO, relay networks, and cognitive radio. More importantly, we have reviewed the fundamental principle of NOMA and pointed out the error propagation phenomenon. Furthermore, we have also considered the channel imperfection and its impact to NOMA performance.

1.2.1.2 Channel Codes

Channel coding is instrumental for achieving higher capacity and reliability. For example, low-density parity-check (LDPC) has been extensively used in 4G, replacing convolutional and turbo codes in previous generations. In 5G NR, polar codes are identified as another promising capacity-achieving coding technique. Polar codes have been adopted in 5G standardization process. For example, 3GPP incorporates polar codes for both uplink and downlink control information communication service, such as enhanced mobile broadband (eMBB), massive machine type communications (mMTC), and ultra-reliable and low latency communications (URLLC). Channel codes for 5G NR should be flexible to support the variable rate and length for both data and control packets. To address that, LDPC has developed several variations, such as quasi-cyclic (QC) LDPC codes for better rate matching and interleaving, as well as parallelism for efficient encoding and decoding [59]; Multi-edge (ME) LDPC mainly for throughput improvement and can scale well in larger block lengths. On the other hand, newly introduced polar code takes advantage of channel polarization, a natural behavior due to signal propagation. Correspondingly, encoding is recursively performed by the channel transformation matrix and creates channels that are either perfectly noiseless or completely noisy. A detailed tutorial of polar codes can be found in [16].

1.2.1.3 Massive MIMO

Massive MIMO refers to applying large-scale antenna elements at transmitter and/or receiver side, usually the number of antenna is hundreds or more. MIMO can exploit spatial diversity or multiplexing, and improve system reliability (for example, lower bit error rate) and throughput, respectively. Compared with legacy MIMO system, massive MIMO brings significant improvements in diversity and multiplexing to fully exploit wireless channel characteristics. One prominent aspect is massive MIMO can generate very narrow beams toward the receiver side. Hence, it can not only increase reception power, but also benefit network capacity and coverage, and ultimately provide better user experience.

These benefits come at a price. Like MIMO, performance gain from massive MIMO largely comes from beamforming and advanced signal processing techniques, which require channel information. If both transmitter and receiver have massive MIMO antennas, their channel is a matrix with hundreds by hundreds of elements. Overhead for accurate channel estimation is prohibitively large. For example, orthogonal pilots are usually applied to obtain channel information; in the case of massive MIMO, maintaining pilot orthogonality is difficult, not to mention practical challenges such as pilot contamination and offset (time and frequency). To address these challenges, prior works have studied robust beamforming design, such that the requirement for accurate channel information can be relaxed. Furthermore, signal processing in massive MIMO is also sophisticated. Traditional optimization methods for throughput maximization or bit error rate (BER) minimization become problematic due to high computation complexity, which hinders the deployment in mobile devices.

It is worth to note that other approaches such as applying out-of-band information, including vision, location, and geometry data to assist beamforming are also studied. Out-of-band information provides complementary details for assisting beamforming steering. These emerging solutions are primarily motivated and enabled by machine learning.

1.2.1.4   Other 5G NR Techniques

5G NR also has other innovations. Recent 3GPP releases 15, 16, and 17 gradually bring more flexibility and enhancement on several aspects. For example, dynamic slot structure caters to different communication needs, for either low-latency or high data-rate application. This structure allows for customized slot design, for examples, adding a longer or shorter cyclic prefix, changing the data frame length, or providing extra guard space. Another innovation is spectrum sharing. In contrast to static database-aided spectrum sharing, which detects secondary users' interference and only allows them to access bands in an opportunistic way, current spectrum sharing is more dynamic, enabled by advanced machine learning-based approach, hence is more efficient and accurate.

1.2.2   Mobile Edge Computing (MEC)

Due to the size, battery, and cost limitations, mobile devices can experience performance bottleneck when computation-intensive tasks are added. More than one decade ago, people solved this problem by introducing the concept of cloud computing. Mobile devices do not perform large-scale computation locally; instead, they send these tasks to remote servers for faster and more secure processing, storage, and sharing. The centralized nature of cloud-based computing can reduce the expenditure cost while providing easier deployment process. However, cloud servers may be located in remote areas, which causes inevitably longer end-to-end transmission and processing delay.

MEC is a new alternative paradigm for the upcoming 5G systems. Instead of transmitting data to the remote servers for processing, MEC provides certain computation capacities locally, for example within the base station in the wireless cellular networks. This paradigm shift can effectively reduce long backhaul latency and energy consumption, as well as support a more flexible infrastructure in a cost-effective way. MEC has attracted extensive research interests recently, not only in the architectural level, but also in specific tasks such as cooperative computation offloading. Computation offloading, which leverages the powerful MEC servers in proximity and sends the computation-intensive tasks for further processing, is a desirable scheme to overcome the physical limitations of user devices (Figure 1.3).

We see this paradigm shift in a more fundamental way. In cloud computing era, even though the data transmission speed is not high, the bottleneck comes mainly from the computation capacity. With Moore's law still being effective, performance of integrated circuit chips grows exponentially. On the other hand, communication technology makes the speed increase almost linearly. Since the goal is to reduce processing speed, it is more beneficial to perform task execution both locally and remotely.

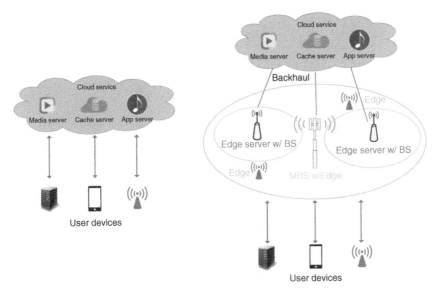

Figure 1.3 Paradigm shift from cloud computing to mobile edge computing.

In order to reduce latency as well as to improve system efficiency, we propose a joint processing scheme in which the total task can be divided into two parts, one for local computing and the other for offloading. To cope with the ever-increasing concerns on energy efficiency, we evaluate the system performance by a new metric, computational efficiency (CE). It is defined as the total number of bits computed with the total energy consumption. The objective is to maximize each user's CE with time constraints (users should finish their task before a required time), energy constraint (each user is powered by battery; hence, the total energy should be below a threshold), and task constraint (each user should finish a minimum number of data bits). Later we show CE is a more appropriate method in terms of finding the balance of more tasks and less energy.

1.2.3 Hybrid and Heterogeneous Communication Architecture for Pervasive IoTs

Recent years have witnessed the unprecedented growth of wearable devices owing to the swift advances in chip design, computing, sensing, and communications technologies. While wearable devices are not new, the past few years have seen a surge in their large-scale use and popularity. A wearable device or simply a wearable refers to a device that can be worn on the body. This rapid rise in popularity was spurred, in part, by technological innovation. Emerging system on chip (SoC) and system in package (SiP) have scaled down the printed circuit board (PCB) size, decreased power consumption, and most importantly, have made it possible to design wearables in a variety of desired shapes (Figure 1.4). Wearable devices provide easier access to information and convenience for their users. They have varying form factors, from low-end health and fitness trackers to high-end virtual reality (VR) devices, augmented reality (AR) helmets, and smart watches. These devices can collect data on heart rates, steps, locations, surrounding buildings, sleeping

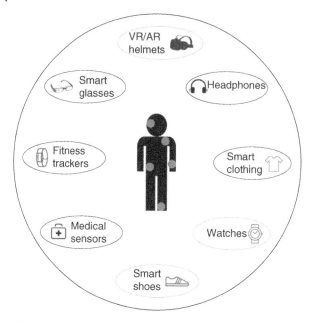

Figure 1.4 Wearable devices may have varying forms, from small medical sensors to entertainment helmets.

cycles, and even brain waves. Yet computing limitations continue to hinder wearables' ability to process data locally. As a result, most devices choose to offload their collected data to other powerful devices or to the clouds. This necessary communication plays a key role in wearable devices. Different applications provided by different wearables may have varying communication requirements. For example, while medical sensors have stringent requirements on latency and reliability, they have a low data rate need. On the other hand, AR/VR devices need both high throughput and low latency for a better user experience.

Wearable devices may not be able to take full advantage of current communication architecture, due to their potential cost and hardware complexity. On the other hand, wearable devices have succeeded in becoming more and more involved in everyday activities requiring voice, image, and video inputs. Human beings are generally sensitive to an approximate 100 ms audible delay and can catch visual delays of less than 10 ms. Furthermore, cell phones and tablets now use primarily touch interaction, a "tactile interaction" that requires a more rigorous delay control, such as 1 ms. A promising heterogeneous and hybrid network architecture is shown in Figure 1.5. It contains small BS (SBS), marco BS (MBS), remote radio head (RRH), and network slice.

1.3 Book Outline

In face of several challenges by 5G and beyond system, this book focuses on technologies that can improve spectral, energy, and computation efficiency. As mentioned above, we

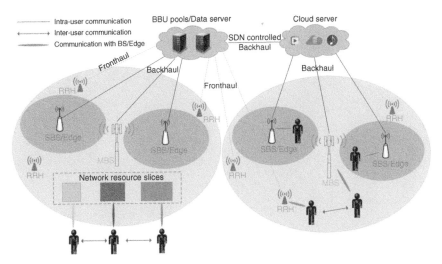

Figure 1.5 A promising network architecture for pervasive IoT communication needs.

mainly study physical layer techniques. Specifically, our first focus (Chapters 1–6) is to provide reader with latest research efforts on 5G NOMA. We have studied NOMA in a systematic way, including applying NOMA to address spectral efficiency and number of connected devices challenges under various network schemes. Our next focus (Chapters 7 and 8) is MEC. MEC is used to reduce computation delay, and we primarily investigate its role for computation offloading. Chapter 9 discusses the emerging wireless paradigm to facilitate distributed machine learning. Chapters 10 and 11 review secure spectrum sharing with machine learning techniques. Lastly, Chapter 12 concludes this book and discusses current and further research directions on 5G and beyond wireless systems.

2

5G Wireless Networks with Underlaid D2D Communications

2.1 Background

2.1.1 MU-MIMO

The narrow beam enabled by massive MIMO brings unprecedented spatial multiplexing. Hundreds of beams can be generated from the BS toward each user, with little interference in the space. This brings new opportunity for multi-user MIMO (MU-MIMO) in the 5G era. MU-MIMO is one type of MIMO technology for wireless communication, in which multiple spatially distributed users with one or more antennas can be transmitted at the same time and frequency by the base station with multiple antennas, at the cost of complicated signal processing and large overhead. Nevertheless, MU-MIMO can greatly improve the system capacity by exploiting the spatial diversity gain among multiple users. The major benefits are improved system throughput (sum rate of users), power, and spectral efficiency. It is expected that massive MIMO will be an important component in 5G cellular networks.

2.1.2 D2D Communication

D2D communication is proposed as another 5G enabler. Compared with traditional BS-centric communication, D2D allows users to initiate and communicate directly, with little or no BS intervention. Intuitively, D2D can reduce signal overhead from/to the BS, lower their transmission power if the receiver is in close proximity, and reduce communication latency. Due to its advantages on power and spectral efficiency, as well as the latency improvement, D2D seeks for larger roles in 5G. However, D2D scheme faces many practical challenges. One is the spectrum coordination. In BS-centric networks, spectrum is allocated by BS, and each user in the system is also synchronized such that they can only use spectrum at their designated slot. D2D is more autonomous and lacks central coordination; hence, spectrum sharing among different users can be problematic. More importantly, D2D will co-exist with cellular users. To ensure cellular users (CUs) performance, usually D2D users (DUs) access spectrum in an opportunistic approach, similar to cognitive networks. The other challenge is QoS. It is difficult to achieve guaranteed QoS,

5G and Beyond Wireless Communication Networks, First Edition. Haijian Sun, Rose Qingyang Hu, and Yi Qian.

primarily due to the lack of centralized resource management. In recent years, BS-assisted D2D emerged. The idea is to allow the BS to assist user and resource coordination, but the level of assistance is less than that in cellular networks.

Nevertheless, to better take advantage of scarce spectrum, DUs can be supported in an underlaid mode, in which they can share the same spectrum with cellular users. In this way, careful interference coordination mechanism is required.

2.1.3 MU-MIMO and D2D in 5G

In recent years, several research have explored the combination of MU-MIMO and D2D in the same system. Early work [100] studied a pair of DUs in the presence of cellular networks, and a resource allocation problem was formulated to maximize network throughput. Later in [204], a more general scheme with multiple pairs of DUs underlying a MU-MIMO cellular network was investigated. With the opportunistic D2D feature, Karakus and Diggavi [93] proved that D2D and MU-MIMO cooperation can boost signal-to-noise ratio (SNR) performance, especially for users at network edge.

As we mentioned in Chapter 1, NOMA can further improve spectral and power efficiency. A more interesting yet challenging question is how to jointly consider NOMA, D2D, and MU-MIMO. The key part of MU-MIMO is to design a suitable precoding matrix for transmitters based on various objective functions, such as overall system capacity or minimum power consumption. When jointly considering MU-MIMO, NOMA, and D2D, a tight coordination among these three mechanisms should be carefully designed so that the overall system performance can be maximized. In [97], NOMA and MU-MIMO are jointly designed to improve the total system throughput. In this chapter, we propose a new mechanism that jointly coordinates beamforming-based MU-MIMO, NOMA, and D2D communications in a downlink cellular network. By supporting DUs in a NOMA/MU-MIMO cellular network, more complicated interference scenarios arise. To address that, we develop two different precoding schemes. One aims to cancel out the BS to DUs interference, while the other one aims to minimize interference among cellular users that coordinate with each other through NOMA and MU-MIMO beamforming. Beamforming is designed together with NOMA pairing and power allocation to significantly improve overall system sum throughput.

Compared with its OMA counterpart, NOMA has a superior performance in terms of spectral efficiency. However, multiuser detection (MUD) is required at the receiver side, which induces a more complex receiver structure and algorithm. In [45], the impact of user pairing on the performance of both fixed power allocation and cognitive radio (CR) inspired NOMA (CR-NOMA) is studied. For the fixed one, NOMA tends to pair users with larger channel gain difference. In [215], a genetic algorithm (GA)-based NOMA pairing in the HetNet is presented, where GA will help reduce the computation workload.

2.2 NOMA-Aided Network with Underlaid D2D

We consider a downlink MU-MIMO cellular network that jointly supports NOMA, MU-MIMO, and underlying DUs. M CUs are randomly distributed, each equipped with

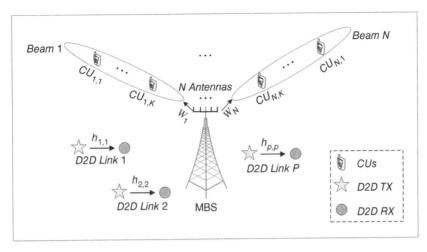

Figure 2.1 System model.

a single antenna [179]. Each BS has N antennas and thus can maximally generate N beamforming vectors. Each beam can support multiple single antenna users by using NOMA, compared with one user in the conventional MU-MIMO system. Furthermore, a total number of P underlaid DU pairs, denoted as $DU_1, DU_2, \ldots, DU_P$, are also randomly distributed, to further exploit current spectrum reuse (Figure 2.1).

For beam n, NOMA allows a set of $\Phi_n = \{u(n, 1), u(n, 2), \ldots, u(n, K)\}$ CUs to be scheduled on the same radio resource simultaneously, $K \geq 2$. We use $u(n, k)$ to denote the CU that is served by beam n with NOMA sequence k in that beam. Assume x_n is the transmitted signal in the n-th beam, and according to NOMA, x_n is a superimposed signal of a total K users in beam n,

$$x_n = \sum_{k=1}^{K} \sqrt{\lambda_{u(n,k)} P_n} s_{u(n,k)}. \tag{2.1}$$

Here $\mathbb{E}(|s_{u(n,k)}|^2) = 1$, $\mathbb{E}(.)$ is the expectation function. $\lambda_{u(n,k)}$ is the fraction of the allocated power to user $u(n, k)$, $\sum_{k=1}^{K} \lambda_{u(n,k)} = 1$. P_n is the total transmitted power for beam n. The total transmission power of a BS is equally partitioned among N beams, i.e. $P_n = \frac{P_{MBS}}{N}$, where P_{MBS} is the total BS transmission power.

At each MBS, a precoding scheme is applied to support MU-MIMO. We denote the precoding matrix as $\mathbf{W}$, which consists of N vectors, i.e.

$$\mathbf{W} = [\mathbf{w}_1, \mathbf{w}_2, \ldots, \mathbf{w}_N], \tag{2.2}$$

where $\mathbf{w}_n \in \mathbb{C}^{N \times 1}$ is the beamforming vector of the n-th beam. The received signals at $u(n, k)$ and DU p can be, respectively, expressed as

$$y_{u(n,k)} = \mathbf{h}_{u(n,k)} \sum_{n=1}^{N} \mathbf{w}_n x_n + \sum_{p=1}^{P} \sqrt{P_D} h_{p,u(n,k)} s_p + n_{u(n,k)}, \tag{2.3}$$

$$y_{DU_p} = \sum_{p'=1}^{P} \sqrt{P_D} h_{p',p} s'_p + \mathbf{h}_p \sum_{n=1}^{N} \mathbf{w}_n x_n + n_p, \tag{2.4}$$

where s_p is the transmitted signal of DU p. We also have $\mathbb{E}(|s_p|^2) = 1$. P_D is the transmission power of DUs. $\mathbf{h}_{u(n,k)}$ and $\mathbf{h}_p$ are the channel gains for downlink CU $u(n,k)$ and for DU p, respectively. $h_{p,u(n,k)}$ is the channel gain between DU p and CU $u(n,k)$, and similarly $h_{p',p}$ is the channel gain between the transmitter of DU p' and the receiver of DU p. We assume the channel information is perfectly now at the BS. $n_{u(n,k)}$ and n_p are i.i.d. additive white Gaussian noise at CU $u(n,k)$ and DU p, respectively. $(n_{u(n,k)}, n_p) \sim \mathcal{CN}(0, 1)$.

2.3 NOMA with SIC and Problem Formation

2.3.1 NOMA with SIC

NOMA is a technique that enables multiple users to share the same spectrum resource simultaneously by employing interference cancellation at the receiver. Within a NOMA group, CU with a weaker channel is normally allocated a higher downlink transmission power so that the strongest received signal within that NOMA group corresponds to the CU with the weakest channel gain in that group. The key idea of SIC is that the received SS is decoded in the ascending order of the respective channel gains or in the descending order of the received signal strength, for all the signals that constitute the SS. The receiver decodes the strongest user signal by treating weaker signals in the SS as interference. The decoded signal can be either the desired signal or can be subtracted from the SS. The decoding process will continue until the receiver successfully decodes its own signal [158].

Channel gains for CUs in the same NOMA group in beam n can be sorted as $|\mathbf{h}_{u(n,1)}| \leq |\mathbf{h}_{u(n,2)}| \leq \cdots \leq |\mathbf{h}_{u(n,K)}|$. Since the decoding order follows the ascending order of channel gains, CU j will decode CU i message, if $i < j$. SIC then removes the decoded message from its observation. CU i treats signals from CUs with index $j > i$ as interference. Assuming perfect interference cancellation, we can rewrite (2.3) as

$$y_{u(n,k)} = \mathbf{h}_{u(n,k)}\mathbf{w}_n\sqrt{\lambda_{u(n,k)}P_n}s_{u(n,k)} + \mathbf{h}_{u(n,k)}\mathbf{w}_n\sum_{k'=1, k'\neq k}^{K}\sqrt{\lambda_{u(n,k')}P_n}s_{u(n,k')}$$

$$+ \mathbf{h}_{u(n,k)}\sum_{n'=1, n'\neq n}^{N}\mathbf{w}_{n'}\sum_{k'=1}^{K}\sqrt{\lambda_{u(n',k')}P_{n'}}s_{u(n',k')} + \sum_{p=1}^{P}\sqrt{P_D}h_{p,u(n,k)}s_p + n_{u(n,k)},$$

where the second term on the right side is the interference from users in the same NOMA group. The third term represents inter-beam interference. After applying SIC, the received signal-to-noise-plus-interference-ratio (SINR) $\gamma_{u(n,k)}$ of CU $u(n,k)$ becomes

$$\gamma_{u(n,k)} = \frac{\lambda_{u(n,k)}P_n|\mathbf{h}_{u(n,k)}\mathbf{w}_n|^2}{I_{u(n,k)}^{N} + I_{u(n,k)}^{U} + I_{u(n,k)}^{D} + \sigma_n^2}, \tag{2.5}$$

where

$$I_{u(n,k)}^{N} = \sum_{k'=k+1}^{K}\lambda_{u(n,k')}P_n|\mathbf{h}_{u(n,k)}\mathbf{w}_n|^2, \tag{2.6}$$

$$I_{u(n,k)}^{U} = \sum_{n'=1, n'\neq n}^{N}P_{n'}|\mathbf{h}_{u(n,k)}\mathbf{w}_{n'}|^2, \tag{2.7}$$

$$I^D_{u(n,k)} = \sum_{p=1}^{P} P_D |h_{p,u(n,k)}|^2, \tag{2.8}$$

respectively represent SIC, inter-beam, and DU interference to CU $u(n,k)$.

Similarly, SINR γ_{DU_p} of the DU p is expressed as

$$\gamma_{DU_p} = \frac{P_D |h_{p,p}|^2}{\sum_{p'=1, p' \neq p}^{P} P_D |h_{p',p}|^2 + \sum_{n=1}^{N} P_n |h_p w_n|^2 + \sigma_n^2}. \tag{2.9}$$

Given SINR, the corresponding user data rate can be calculated as $f(\mathbb{E}\{\gamma\})$ by using Shannon capacity formula,

$$f(\mathbb{E}\{\gamma\}) = \log(1 + \mathbb{E}\{\gamma\}). \tag{2.10}$$

Here we normalize the bandwidth at MBS to 1.

2.3.2 Problem Formation

The design objective is to maximize the total system sum throughput from both CUs and DUs. To this end, we need to determine (i) the NOMA set of each beam, i.e. Φ_n; (ii) the power allocation factor $\lambda_{u(n,k)}$ for each user k in the NOMA set of beam n; and (iii) the precoding vector w_n. Therefore, the problem can be formulated as follows.

$$\max_{\Phi_n, w_n, \lambda_{u(n,k)}} \sum_{n=1}^{N} \sum_{k=1}^{K} f(\mathbb{E}\{\gamma_{u(n,k)}\}) + \sum_{p=1}^{P} f(\mathbb{E}\{\gamma_{DU_p}\}) \tag{2.11}$$

subject to

$$\sum_{k=1}^{K} \lambda_{u(n,k)} = 1, \ n = 1, 2, \dots, N, \tag{2.12}$$

$$f(\mathbb{E}\{\gamma_{u(n,k)}\}) > R_0, \ \forall k \neq K, \tag{2.13}$$

$$w_n \in \mathbb{C}^{N \times 1}. \tag{2.14}$$

Constraint (2.12) is the summation of user power in one beam. Constraint (2.13) sets a lower rate limit for users that experience SIC interference in NOMA to ensure good user experience. $\gamma_{u(n,k)}$ and γ_{DU_p} are rates calculated based on (2.5) and (2.9), respectively. The optimization problem is a non-convex problem that needs to determine $\Phi_n, w_n, \lambda_{u(n,k)}$ jointly. To make this problem feasible to solve, in Section 2.4, we seek a heuristic solution by decomposing the original problem into two sub-problems. We first develop different precoding methods, which aim to suppress either the inter-beam interference among CUs or the interference from CUs to DUs. Based on the precoding matrices, we further define a user grouping and power allocation algorithm for NOMA.

2.4 Precoding and User Grouping Algorithm

In this section, we first construct a beamforming vector w_n for each beam that can effectively reduce or eliminate some interferences. Based on the selected precoding scheme, we further solve the user grouping and power allocation problem, in order to maximize the total system throughput.

2.4.1 Zero-Forcing Beamforming

Normally the number of transmit antennas n_T should be larger than or equal to the number of receiver antennas n_R, i.e. $n_T \geq n_R$, so that the transmitter side will have enough degree of freedom to generate a precoding matrix that can effectively eliminate the inter-user interference. In this chapter, each MBS has N transmit antennas and can generate N beams. Within each beam, $K(K \geq 2)$ users can be supported by using NOMA. Thus, the total number of receive antennas in this case is $N \times K$, which is larger than N. Existing literatures have observed and addressed this issue. In [167], a coordinated transmit-receive block diagonalization algorithm is put forward. However, the receive antenna set employs a joint precoding matrix, which requires information exchange among different users and consequently adds extra complexity. Here, we consider two zero-forcing precoding methods. The first one aims to minimize the inter-beam interference for CUs, while the second one aims to eliminate the interference from MBS to DUs.

2.4.1.1 First ZF Precoding

In this scheme, we first select one user from each beam and then generate the beamforming matrix based on N selected users. Specifically, users with the largest channel gain in each beam are selected. The channel gain vector for these N selected CUs is denoted as $\mathbf{H} = [\mathbf{h}_{u(1,K)}, \mathbf{h}_{u(2,K)} \dots \mathbf{h}_{u(N,K)}]$. The zero-forcing beamforming vector is calculated based on:

$$\mathbf{h}_{u(n,K)}\mathbf{w}_m = 0, \text{ if } m \neq n. \tag{2.15}$$

Thus, $\mathbf{w}_m$ should lie in the null space of $\tilde{\mathbf{H}}_n$ [167]. Here, $\tilde{\mathbf{H}}_n$ is defined as

$$\tilde{\mathbf{H}}_n = [\mathbf{h}_{u(1,K)}, \dots, \mathbf{h}_{u(n-1,K)}, \mathbf{h}_{u(n+1,K)}, \dots, \mathbf{h}_{u(N,K)}], \tag{2.16}$$

which consists of downlink channel vectors for CUs from all beams except from beam n.

2.4.1.2 Second ZF Precoding

The first zero forcing (ZF)-based method helps reduce inter-beam interference $I^U_{u(n,K)} = 0$ in (2.5). Since we aim to maximize the total sum rate in the system, the total throughput from DUs contributes to the total throughput as well. Therefore, the second precoding method helps reduce the interference between CUs and DUs, i.e. $\sum_{n=1}^{N} P_n |\mathbf{h}_p \mathbf{w}_n|^2 = 0$ in (2.9). Hence we should set $\mathbf{h}_p \mathbf{w}_n = 0$, for all n. Or equivalently,

$$\mathbf{w}_n = null(\mathbf{H}_D), \tag{2.17}$$

where $\mathbf{H}_D = [\mathbf{h}_1, \dots, \mathbf{h}_P]$, and $null(.)$ is the null space or kernel of a matrix.

2.4.2 User Grouping and Optimal Power Allocation

After the beamforming vector is determined, we need to group NOMA users into each beam and further decide power allocation for CUs within each NOMA group. One way is to do an exhaustive search, but the complexity will grow exponentially with N. Inspired by Ding *et al.* [45] and Kimy *et al.* [97], NOMA would prefer to group users with greater channel differences. On the other hand, precoding matrix $\mathbf{W}$ is designed to minimize inter-beam interference or CU to DU interference. When combining NOMA and precoding, NOMA groups users with highly correlated channels so that using the precoding matrix generated by the representative CU in each beam can achieve a small inter-beam or CU-DU interference. Therefore, the criterion for NOMA user grouping is to choose CUs with highly correlated channels but with big channel gain differences in each beam. For simplicity, we set $K = 2$. In each NOMA pair, we denote the user with a weaker channel gain as the first user while the stronger one as the second user.

2.4.2.1 First ZF Precoding

Since the beamforming matrix is designed based on the null space of the second users in all N beams, second users will not receive any inter-beam interference. Thus their SINR is

$$\gamma_{u(n,2)} = \frac{\lambda_{u(n,2)}P_n|\mathbf{h}_{u(n,2)}|^2}{I^D_{u(n,2)} + \sigma_n^2}. \tag{2.18}$$

The first users, on the other hand, will receive non-zero inter-beam interference as the precoded signals from other beams will have components projected into the first user signal space. Their SINR is expressed as

$$\gamma_{u(n,1)} = \frac{(1 - \lambda_{u(n,2)})P_n|\mathbf{h}_{u(n,1)}\mathbf{w}_n|^2}{|\mathbf{h}_{u(n,1)}\mathbf{w}_n|^2\lambda_{u(n,2)}P_n + I^D_{u(n,1)} + I^U_{u(n,1)} + \sigma_n^2}. \tag{2.19}$$

The optimal power allocation factor $\lambda_{u(n,2)}$ is yet to be solved. Based on the optimization problem proposed in Section 2.3, we form a new problem that aims to maximize the sum capacity in each beam.

$$\max_{\lambda_{u(n,2)}} \sum_{k=1}^{2} f(\mathbb{E}\{\gamma_{u(n,k)}\}) \tag{2.20}$$

subject to

$$0 < \lambda_{u(n,2)} < 1, \tag{2.21}$$

$$f(\mathbb{E}\{\gamma_{u(n,1)}\}) \geq R_0. \tag{2.22}$$

The problem defined above is convex with respect to $\lambda_{u(n,2)}$ and its Karush–Kuhn–Tucker (KKT) conditions are given as follows.

$$\frac{\partial\left(\sum_{k=1}^{2} f(\mathbb{E}\{\gamma_{u(n,k)}\})\right)}{\partial\lambda^*_{u(n,2)}} = \mu\frac{\partial\left(R_0 - f(\mathbb{E}\{\gamma_{u(n,1)}\})\right)}{\partial\lambda^*_{u(n,2)}}, \tag{2.23}$$

$$R_0 - f(\mathbb{E}\{\gamma_{u(n,1)}\})|_{\lambda^*_{u(n,2)}} \leq 0, \tag{2.24}$$

$$\mu \geq 0, \tag{2.25}$$

$$\mu\left(R_0 - f(\mathbb{E}\{\gamma_{u(n,1)}\})|_{\lambda^*_{u(n,2)}}\right) = 0. \tag{2.26}$$

Equation (2.23) is the stationarity condition and μ is KKT multiplier, (2.24) is the primal feasibility, (2.25) is dual feasibility, and (2.26) is the complementary slackness.

Solving for (2.23), we can get

$$\lambda^*_{u(n,2)} = \frac{(\mathcal{I}_{D2} + 1)\left((\mathcal{I}_{D1} + 1 + \Sigma)\mathcal{H}_2 - (1 + \mu)\mathcal{H}_1\right)}{\mathcal{H}_1\mathcal{H}_2\rho(\mu - \mathcal{I}_{D2})}, \tag{2.27}$$

where, $\rho = P_n/\sigma_n^2$ is the transmit SNR, $\mathcal{H}_2 = |\mathbf{h}_{u(n,2)}|^2$, $\mathcal{H}_1 = |\mathbf{h}_{u(n,1)}\mathbf{w}_n|^2$ is the channel gain for users 2 and 1, $\mathcal{I}_{D1} = I^D_{u(n,1)}/\sigma_n^2$, $\mathcal{I}_{D2} = I^D_{u(n,2)}/\sigma_n^2$ is the interference-to-noise ratio of users 1 and 2, respectively. $\Sigma = I^U_{u(n,1)}/\sigma_n^2$ is the inter-beam interference-to-noise ratio.

Clearly, $\mu \neq 0$. Otherwise, $\lambda^*_{u(n,2)} < 0$ cannot satisfy (2.21). Therefore, we can solve (2.26) for the optimal $\lambda^*_{u(n,2)}$,

$$\lambda^*_{u(n,2)} = \frac{\rho\mathcal{H}_1 + \mathcal{I}_{D1} + 1 + \Sigma}{2^{R_0}\rho\mathcal{H}_1} - \frac{\mathcal{I}_{D1} + 1 + \Sigma}{\rho\mathcal{H}_1},$$ (2.28)

$$\lambda^*_{u(n,1)} = 1 - \lambda^*_{u(n,2)}.$$ (2.29)

2.4.2.2 Second ZF Precoding

In the second ZF precoding, the inter-beam interference remains for both first and second users. Their respective SINR are as follows.

$$\gamma_{u(n,2)} = \frac{\lambda_2\rho\mathcal{H}'_2}{\Sigma'_2 + I'_{D2} + 1},$$ (2.30)

$$\gamma_{u(n,1)} = \frac{(1 - \lambda_2)\rho\mathcal{H}'_1}{\lambda_2\rho\mathcal{H}'_1 + \Sigma'_1 + I'_{D1} + 1}.$$ (2.31)

Similarly, λ_2 is the power allocation factor for the user with stronger channel. $\mathcal{H}'_1 = |\mathbf{h}_{u(n,1)}\mathbf{w}_{ZF2}|^2$, $\mathcal{H}'_2 = |\mathbf{h}_{u(n,2)}\mathbf{w}_{ZF2}|^2$ are the channel gains for users 1 and 2, respectively. $\Sigma'_2 = \sum_{n'=1,n'\neq n}^{N} \rho|\mathbf{h}_{u(n,2)}\mathbf{w}_{ZF2}|^2$ and $\Sigma'_1 = \sum_{n'=1,n'\neq n}^{N} \rho|\mathbf{h}_{u(n,1)}\mathbf{w}_{ZF2}|^2$. I'_{D1} has the same format as $\mathcal{I}_{D1}$ but with different precoding vector, the same to I'_{D2}. We form a similar optimization problem as in (2.20) and detailed derivations are omitted here. The respective optimal power allocation factor for the second and first users is

$$\lambda^*_2 = \frac{\rho\mathcal{H}'_1 + I'_{D1} + 1 + \Sigma'_1}{2^{R_0}\rho\mathcal{H}'_1} - \frac{I'_{D1} + 1 + \Sigma'_1}{\rho\mathcal{H}'_1},$$ (2.32)

$$\lambda^*_1 = 1 - \lambda^*_2.$$ (2.33)

2.5 Numerical Results

In this section, we present the performance results from simulation. The coverage area of MBS is circular with a radius of 500 m. The number of transmit antennas is $N = 3$. The total numbers of CUs and DUs are $M = [8, 16, 32, 60, 90]$ and $P = 2$, respectively. M varies in order to study the multi-user diversity effect. The distance with each DU pair is fixed at 30 m. The wireless channel consists of pathloss, shadowing, and Rayleigh fading with a pathloss exponent 2. P_{MBS} and P_D are set to 30 Watt and 1 Watt, respectively.

For comparison purpose, instead of using NOMA in each beam, we apply a traditional TDMA scheme here to support these 2 users in each beam. Specifically, we allocate an equal number of time slots to 2 TDMA users. The scheme is also referred as "Naive TDMA."

$$R_{TDMA} = \frac{1}{2}\left(\log(1 + \gamma_1) + \log(1 + \gamma_2)\right).$$ (2.34)

Figure 2.2 presents the system capacity of two proposed ZF precoding methods as the number of users grows, and the results are scaled over the highest achievable rate. Here we

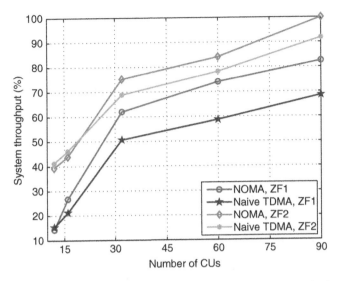

Figure 2.2 System capacity of two proposed ZF precoding methods vs. TDMA as the number of user grows ($R_0 = 0.5$ bit/s/Hz).

set $R_0 = 0.5$ bit/s/Hz. We can see that NOMA outperforms naive TDMA in both precoding schemes when the number of CUs is large. However, when the number is small, limited number of CUs can be chosen to perform NOMA; thus, the performance gain is not obvious, even worse than TDMA. We also find that using ZF2 leads to a higher overall system throughput than ZF1. Because with ZF2, DUs experience a much lower interference than with ZF1 so that the throughput elevation from DUs exceeds the throughput degradation from CUs due to inter-beam interference, which results a net gain on overall system throughput. Moreover, as the user number increases, the system benefits more from NOMA+MU-MIMO due to a higher multiuser diversity gain.

DUs normally are considered as a complementary communication method. So we are particularly interested in the performance of CUs. In Figure 2.3, the throughput of CUs is calculated. NOMA shows a superior spectral efficiency compared with naive TDMA. In this case, ZF1 has a much better performance than ZF2 since ZF1 precoding eliminates inter-beam interference for CUs, while ZF2 aims to eliminate interference from CUs to DUs. But if we combine results from both Figures 2.2 and 2.3, we can see that the overall throughput is higher with ZF2 since DUs are configured with a very good channel setting so that they contribute to overall throughput significantly.

2.6 Summary

In this chapter, we study the performance of a cellular network that supports NOMA, MU-MIMO, and D2D communications. Specifically, we use NOMA and MU-MIMO for the cellular downlink users to improve overall system spectrum efficiency. D2D users are further supported in the underlay mode to exploit the frequency reuse again. Two different precoding mechanisms are defined. We formulate an optimization problem

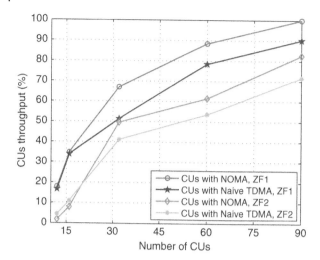

Figure 2.3 CUs capacity of two proposed ZF precoding methods vs. TDMA as the number of user grows ($R_0 = 0.5$ bit/s/Hz).

aiming to maximize the system performance and develop a suboptimal approach to solve the problem in two steps. Simulation results show NOMA and MU-MIMO altogether will improve the overall cellular user throughput significantly. When underlay D2D users are added, two different precoding schemes lead to different performance, with one favoring CUEs and one favoring DUEs. But both lead to a net system gain.

3

5G NOMA-Enabled Wireless Networks

3.1 Background

In Chapter 2, we incorporated NOMA with MU-MIMO in a D2D underlaid system. The advantages of applying NOMA in such a scheme are in two folds: (i) it can support more users simultaneously; (ii) overall system performance in terms of total throughput is also improved. Intuitively, NOMA will create a win-win situation for a pair of users with strong and weak channel condition. The reason is that the stronger user is typically bandwidth-limited, while the weaker user is interference-limited. In NOMA, signals for both users are set to transmit simultaneously, so the bandwidth-limited user can get more spectrum resources, while the interference-limited user can obtain a larger portion of power. This will benefit the whole system in terms of fairness and throughput.

In [214], the concept of NOMA is discussed from the information theoretic perspective, and the conclusion is that NOMA can have a better performance compared with OMA in terms of both system sum rate and user individual rate, especially when the users channel gains are distinct. In [171] and [170], a similar downlink MIMO and NOMA system model is proposed, and the authors solved the optimization problem with bisection power search algorithm and applied the singular value decomposition (SVD) if the channel state information (CSI) is available at the BS, or equally distribute powers among different antennas if CSI is unknown for the precoding design.

These works, however, all assume a perfect subtraction of previous user signals in SIC such that there's no residual interference which will affect the current decoding. This assumption turns out to be a strong one since various factors can actually cause errors, such as deep fading, imperfect decoding, and channel estimation errors [140]. In the case of decoding more users' signal, errors from previous will accumulate and greatly affect the next stage (we refer this as error propagation). In this chapter, we take error propagation into consideration, a concept that already exists in code division multiple access (CDMA) systems. Similar papers can be found in [8] and [9]. In fact, CDMA shares some common features with NOMA. Both of them exploit the multiuser interference to achieve a higher performance rather than simply avoid it. Performance gain also largely depends on some assumptions like perfect channel estimation and power allocation, and violations can cause

5G and Beyond Wireless Communication Networks, First Edition. Haijian Sun, Rose Qingyang Hu, and Yi Qian.
© 2024 John Wiley & Sons Ltd. Published 2024 by John Wiley & Sons Ltd.

serious performance degradation. In this chapter, we propose a general error propagation model in a downlink MIMO NOMA system, where decoding errors are modeled as residual interference. An optimization problem is formulated to maximize the total data rate of two users.

3.2 Error Propagation in NOMA

We consider a downlink wireless communication which jointly supports NOMA and MU-MIMO. In the system, a BS with power P_{BS} is equipped with M antennas. Two user equipments (UEs) are randomly deployed in this area, each has N antennas [172].

Due to the usage of NOMA, two UEs can receive signals from the BS simultaneously. Besides, the BS is assumed to have an accurate CSI of UEs based on training sequences and feedback mechanism. We denote $\mathbf{H}_k$ and $\mathbf{H}_n$ (both with dimension $\mathbb{C}^{N \times M}$) as the channel gain matrix of UE k and UE n, respectively. h_{ij} is the element from ith row and jth column in the matrix, and it is modeled as the product of large-scale path loss and fading, i.e. $h_{ij} = l_{ij}^{-\alpha} h_0$, where l_{ij} is the distance between UE and BS, α is the path loss exponent, and h_0 is the Gaussian random variable with distribution $h_0 \sim \mathcal{CN}(0, 1)$.

The transmitted signal from the BS is:

$$\mathbf{x}_{BS} = \mathbf{W}_n \mathbf{x}_n + \mathbf{W}_k \mathbf{x}_k, \tag{3.1}$$

where $\mathbf{W}_n$ and $\mathbf{W}_k$ are precoding matrices with dimension $\mathbb{C}^{M \times N}$, $\mathbf{x}_n$ and $\mathbf{x}_k \in \mathbb{C}^{N \times 1}$ are messages for UE n and UE k, respectively. $\mathbb{E}(\mathbf{x}_n \mathbf{x}_n^H) = \mathbb{E}(\mathbf{x}_n \mathbf{x}_n^H) = \mathbf{I}_N$, $\mathbb{E}(.)$ is the expectation function and $\mathbf{I}_N$ is a $N \times N$ identity matrix.

The received signal at UE n is

$$\mathbf{y}_n = \mathbf{H}_n \mathbf{x}_{BS} + \mathbf{n}_n. \tag{3.2}$$

Similarly, UE k will receive,

$$\mathbf{y}_k = \mathbf{H}_k \mathbf{x}_{BS} + \mathbf{n}_k, \tag{3.3}$$

where $\mathbf{n}_i, i = \{n, k\}$ is the i.i.d additive gaussian noise which follows $\mathcal{CN}(0, \sigma^2 \mathbf{I}_N)$.

3.3 SIC and Problem Formulation

The key idea of SIC can be summarized as a process of decoding, reconstruction, and subtraction (DRS). Upon the reception of the composite signal, DRS will start with the strongest user signal and treat others as interference. After the successful decoding, the data will re-encode based on user channel estimation and constellation. The reconstructed signal should be fairly close to the received signal if everything is perfect. Then, the user will subtract this signal from the aggregated signal so that the next DRS will see less interference if the intended message is not decoded [158]. However, DRS can be affected by error propagation, and we will show this concept below.

3.3.1 SIC with Error Propagation

Sequential decoding can be affected by error propagation. Consider a simpler system with one BS and two UEs, in which UE 1 and UE 2 form a NOMA pair. The power of the BS is P and the channel gains for UE 1 and UE 2 are h_1 and h_2, respectively. Without loss of generality, let $h_1 > h_2$. The transmitted signal can be expressed as

$$x_t = \sqrt{\theta_1 P s_1} + \sqrt{\theta_2 P s_2}, \tag{3.4}$$

where θ_1 and θ_2 are power allocation factors, $\theta_1 < \theta_2$ for QoS consideration, and $\theta_1 + \theta_2 = 1$. s_1 and s_2 are normalized signals.

At the receiver side, UE 1 will get $y_1 = h_1 x_t + n_1 = h_1(\sqrt{\theta_1 P s_1} + \sqrt{\theta_2 P s_2}) + n_1$. Clearly the signal for UE 2 has a larger power than that for UE 1, thus, at the first stage, UE 1 will decode UE 2's signal. Let $R_{1,2}$ denote the achievable data rate for UE 1 to detect UE 2's message, it can be expressed as,

$$R_{1,2} = \log_2\left(1 + \frac{\theta_2 P |h_1|^2}{\theta_1 P |h_1|^2 + n_1^2}\right). \tag{3.5}$$

UE 1 then reconstructs this message according to a prior known constellation and channel gain. After that UE 1 will subtract UE 2's signal and decode its own, and the data rate is given by,

$$R_1 = \log_2\left(1 + \frac{\theta_1 P |h_1|^2}{n_1^2}\right). \tag{3.6}$$

The received signal for UE 2 is $y_2 = h_2 x_t + n_2 = h_2(\sqrt{\theta_1 P s_1} + \sqrt{\theta_2 P s_2}) + n_2$, since the desired signal has a larger power, so it can be detected directly. The achievable data rate for UE 2 is simply

$$R_2 = \log_2\left(1 + \frac{\theta_2 P |h_2|^2}{\theta_1 P |h_2|^2 + n_2^2}\right). \tag{3.7}$$

The above procedure, however, depends on the perfect DRS of UE 2's signal at UE 1, which is a strong assumption, since various factors such as deep fading can affect the signal detection and decoding. Assuming at UE 1 side, the DRS procedure is not perfect, there will be residual signal power at stage 2 when decoding its own message. As a result, the data rate for UE 1 becomes,

$$R_1' = \log_2\left(1 + \frac{\theta_1 P |h_1|^2}{\beta \theta_2 P |h_1|^2 + n_1^2}\right), \tag{3.8}$$

where β is the error propagation factor, which is inversely proportional to the signal-to-noise-plus-interference-ratio (SINR) of (3.5), i.e. $\beta \propto \frac{\theta_1 P |h_1|^2 + n_1^2}{\theta_2 P |h_1|^2}$ and $0 \leq \beta \leq 1$. $\beta = 0$ represents the perfect decoding, which is the same as (3.6). While $\beta = 1$ is the worst case that the DRS of UE 2 is totally unsuccessful and UE 1 has to treat its entire signal as interference. (In this case, it has the same result as without SIC.)

In our system model, if we assume $H_n H_n^H \succ H_k H_k^H$, here $\succ$ means if $A \succ B$, then $(A - B)$ is a positive definite matrix. This assumption implies UE n has a better channel condition and hence can decode UE k's message. Thus, at UE n, we have

$$R_{n,k} = \log_2 \det \left(\mathbf{I} + (\sigma^2 \mathbf{I} + \mathbf{H}_n \mathbf{W}_n \mathbf{W}_n^H \mathbf{H}_n^H)^{-1} \mathbf{H}_n \mathbf{W}_k \mathbf{W}_k^H \mathbf{H}_n^H \right), \tag{3.9}$$

which is the maximum achievable rate for UE k at UE n. Considering the error propagation, the data rate for UE n's own message would be,

$$R_n = \log_2 \det \left(\mathbf{I} + (\sigma^2 \mathbf{I} + \beta \mathbf{H}_n \mathbf{W}_k \mathbf{W}_k^H \mathbf{H}_n^H)^{-1} \mathbf{H}_n \mathbf{W}_n \mathbf{W}_n^H \mathbf{H}_n^H \right). \tag{3.10}$$

The error propagation factor β is assumed be a fixed value.

While at UE k, the desired signal can be decoded directly.

$$R_{k,k} = \log_2 \det \left(\mathbf{I} + (\sigma^2 \mathbf{I} + \mathbf{H}_k \mathbf{W}_n \mathbf{W}_n^H \mathbf{H}_k^H)^{-1} \mathbf{H}_k \mathbf{W}_k \mathbf{W}_k^H \mathbf{H}_k^H \right). \tag{3.11}$$

In order for UE k to have a fairly small BER, the maximum allowable data rate for UE k is,

$$R_k = \min \{R_{n,k}, R_{k,k}\}. \tag{3.12}$$

Here we normalize the bandwidth at the BS to 1.

Next, we show that $R_k = R_{k,k}$, the proof follows appendix A in [171] and can be briefly summarized as follows.

Proof: Since $\mathbf{H}_n \mathbf{H}_n^H > \mathbf{H}_k \mathbf{H}_k^H$, we can write $\mathbf{H}_n = \mathbf{M} \mathbf{H}_k$, where $\mathbf{M}$ is a $N \times N$ matrix and $\mathbf{M} \mathbf{M}^H > \mathbf{I}_N$.

Due to the property of determinant operation, we can rewrite $R_{n,k}$ as

$$R_{n,k} = \log_2 \det \left(\mathbf{I} + \mathbf{W}_k^H \mathbf{H}_n^H (\sigma^2 \mathbf{I} + \mathbf{H}_n \mathbf{W}_n \mathbf{W}_n^H \mathbf{H}_n^H)^{-1} \mathbf{H}_n \mathbf{W}_k \right) \tag{3.13}$$

Define $\mathbf{Q}_{n,k} = \mathbf{W}_k^H \mathbf{H}_n^H (\sigma^2 \mathbf{I} + \mathbf{H}_n \mathbf{W}_n \mathbf{W}_n^H \mathbf{H}_n^H)^{-1} \mathbf{H}_n \mathbf{W}_k$ and $\mathbf{Q}_{k,k} = \mathbf{W}_k^H \mathbf{H}_k^H (\sigma^2 \mathbf{I} + \mathbf{H}_k \mathbf{W}_n \mathbf{W}_n^H \mathbf{H}_k^H)^{-1} \mathbf{H}_k \mathbf{W}_k$, then we substitute $\mathbf{H}_n = \mathbf{M} \mathbf{H}_k$ in $\mathbf{Q}_{n,k}$.

$$\begin{aligned} \mathbf{Q}_{n,k} &= \mathbf{W}_k^H \mathbf{H}_k^H (\sigma^2 \mathbf{I} + \mathbf{M} \mathbf{H}_k \mathbf{W}_n \mathbf{W}_n^H \mathbf{H}_k^H \mathbf{M}^H)^{-1} \mathbf{H}_n \mathbf{W}_k \\ &= \mathbf{W}_k^H \mathbf{H}_k^H (\sigma^2 (\mathbf{M}^H \mathbf{M})^{-1} + \mathbf{H}_k \mathbf{W}_n \mathbf{W}_n^H \mathbf{H}_k^H)^{-1} \mathbf{H}_k \mathbf{W}_k \\ &> \mathbf{W}_k^H \mathbf{H}_k^H (\sigma^2 \mathbf{I} + \mathbf{H}_k \mathbf{W}_n \mathbf{W}_n^H \mathbf{H}_k^H)^{-1} \mathbf{H}_k \mathbf{W}_k \\ &= \mathbf{Q}_{k,k} \end{aligned} \tag{3.14}$$

Thus, $\log_2 \det (\mathbf{I} + \mathbf{Q}_{n,k}) > \log_2 \det (\mathbf{I} + \mathbf{Q}_{k,k})$, which means $R_{n,k} > R_{k,k}$, so $R_k = R_{k,k}$.

3.3.2 Problem Formation

In this chapter, we intend to maximize the system throughput by applying NOMA and MU-MIMO. The problem can be formed as following.

$$\max_{\mathbf{W}_n, \mathbf{W}_k} (R_n + R_k) \tag{3.15a}$$

$$\text{tr}(\mathbf{W}_n \mathbf{W}_n^H + \mathbf{W}_k \mathbf{W}_k^H) \leq P_{BS}, \tag{3.15b}$$

$$R_k \geq R_0. \tag{3.15c}$$

Equation (3.15b) is the constraint for maximum allowed power from the BS. 3.15c sets a minimum data rate for the weaker UE. R_n and R_k can be calculated based on (3.10) and (3.11), respectively. One note here is due to the error propagation, the stronger UE may suffer severe residual interference from the weaker one; thus, its data rate may be lower. However, we do not consider this situation in the chapter; the lower data rate limit is only for the weaker UE.

When a resource block (RB) is available, the BS needs to determine the following: (i) How to properly design the precoding matrix; (ii) How to allocate the power to each UE.

The above optimization problem is hard to solve, and the reason is that it imposes the error propagation, which makes the utility function 3.15a hard to track. Besides, when calculating R_i, we also need to determine the precoding matrix $\mathbf{W}_i$, for $i = n, k$. In Section 3.4, we propose a unified precoding matrix formation algorithm, and then we focus on the power allocation with residual interference.

3.4 Precoding and Power Allocation

3.4.1 Precoding Design

Let $\mathrm{tr}(\mathbf{W}_n \mathbf{W}_n^H) = P_n$ and $\mathrm{tr}(\mathbf{W}_k \mathbf{W}_k^H) = P_k$. The optimization problem can be revised as:

$$\max_{\mathbf{W}_n, \mathbf{W}_k} (R_n + R_k) \tag{3.16}$$

subject to

$$\mathrm{tr}(\mathbf{W}_n \mathbf{W}_n^H) = P_n, \tag{3.17}$$

$$\mathrm{tr}(\mathbf{W}_k \mathbf{W}_k^H) \leq P_{BS} - P_n, \tag{3.18}$$

$$R_k \geq R_0. \tag{3.19}$$

The introduced error propagation model increases the complexity of the optimization problem. Basically it is a MIMO broadcast channel (BC) in the downlink. So it can be converted to multiple access channel (MAC) in the uplink using BC-MAC duality. But it requires extensive matrix calculation and is not easy to solve. Here in this chapter, we introduce the equivalent channel and its respective precoding solution.

From (3.10), we denote $\mathbf{H}n_{eq}$ as the equivalent channel of UE n and it can be expressed as $(\sigma^2 \mathbf{I} + \beta \mathbf{H}_n \mathbf{W}_k \mathbf{W}_k^H \mathbf{H}_n^H)^{-1/2} \mathbf{H}_n$. We then rewrite (3.10) in terms of the equivalent channel $\mathbf{H}n_{eq}$.

$$R_n = \log_2 \det (\mathbf{I} + \mathbf{H}n_{eq} \mathbf{W}_n \mathbf{W}_n^H \mathbf{H}n_{eq}^H). \tag{3.20}$$

Similarly, (3.11) can be expressed as,

$$R_k = \log_2 \det (\mathbf{I} + \mathbf{H}k_{eq} \mathbf{W}_k \mathbf{W}_k^H \mathbf{H}k_{eq}^H), \tag{3.21}$$

where $\mathbf{H}k_{eq} = (\sigma^2 \mathbf{I} + \mathbf{H}_k \mathbf{W}_n \mathbf{W}_n^H \mathbf{H}_k^H)^{-1/2} \mathbf{H}_k$.

Thus, we can treat the problem as two point-to-point MIMO UEs with a total power constraint, which is already well known in the literature [63]. However, due to the imposed minimum data rate requirement for UE k, It is not necessarily the optimal solution. The suboptimal precoding can be formed as follows. First, take the SVD of the equivalent channel,

$$\mathbf{U}_n \mathbf{E}_n \mathbf{U}_n^H = \mathbf{H}n_{eq}^H \mathbf{H}n_{eq}, \tag{3.22}$$

where $\mathbf{U}_n$ is the unitary matrix and its columns are a set of orthonormal eigenvectors of $Hn_{eq}^H Hn_{eq}$, $\mathbf{E}_n$ is a diagonal matrix. Therefore, the precoding matrix can be formed as,

$$\mathbf{W}_n \mathbf{W}_n^H = \mathbf{U}_n \tilde{\mathbf{E}}_n \mathbf{U}_n^H, \tag{3.23}$$

where $\tilde{\mathbf{E}}_n$ is calculated from water-filling process with respect to the elements in the diagonal matrix $\mathbf{E}_n$, i.e. $\tilde{\mathbf{E}}_n = \lceil \lambda_n \mathbf{I} - (\mathbf{E}_n)^{-1} \rceil^+$, here λ_n is a parameter to ensure the power constraint $\mathrm{tr}(\mathbf{W}_n \mathbf{W}_n^H) = P_n$, and $\lceil a \rceil^+ = \max(a, 0)$.

The precoding matrix for UE k can be formed in the same way. However, two problems remain here: (i) The calculation of $\mathbf{W}_n$ involves $\mathbf{W}_k$ and vice versa; (ii) Power P_n and P_k are unknown. Next, we propose an iterative way to solve for precoding generation under the assumption that each UE's power is known as *a prior*.

We start with $\mathbf{W}_k \mathbf{W}_k^H = \frac{P_k}{M} \mathbf{I}_M$, and calculate Hn_{eq}, $\mathbf{W}_n \mathbf{W}_n^H$ and Hk_{eq} sequentially, then update $\mathbf{W}_k \mathbf{W}_k^H$ according to the new Hk_{eq}. The process will continue until it reaches the maximum iteration number. To make further clarification, the algorithm for precoding design is summarized in **Algorithm 3.1**.

Algorithm 3.1 Iterative Precoding Design

1: **Initialization:** Given power P_n and P_k, maximum iteration number *MAXITER*.
2: $\mathbf{W}_k \mathbf{W}_k^H = \frac{P_k}{M} \mathbf{I}_M$.
3: **for** $i = 1$ to *MAXITER* **do**
4:   Calculate Hn_{eq} based on $\mathbf{W}_k \mathbf{w}_k^H$
5:   Solve for $\mathbf{W}_n \mathbf{W}_n^H$ from the SVD of $Hn_{eq}^H Hn_{eq}$.
6:   Calculate Hk_{eq} based on $\mathbf{W}_n \mathbf{W}_n^H$
7:   Update $\mathbf{W}_k \mathbf{W}_k^H$ from the SVD of $Hk_{eq}^H Hk_{eq}$.
8: **end for**
9: Output R_n, R_k, $\mathbf{W}_k \mathbf{W}_k^H$ and $\mathbf{W}_n \mathbf{W}_n^H$.

A note here is that the covariance matrix $\mathbf{W}_k \mathbf{W}_k^H$ and $\mathbf{W}_n \mathbf{W}_n^H$ actually characterize the data rate, not $\mathbf{W}_n$ or $\mathbf{W}_k$ individually. And an easy way to find $\mathbf{W}_n$ and $\mathbf{W}_k$ is,

$$\mathbf{W}_n = \mathbf{U}_n \tilde{\mathbf{E}}_n^{\frac{1}{2}}, \mathbf{W}_k = \mathbf{U}_k \tilde{\mathbf{E}}_k^{\frac{1}{2}}, \tag{3.24}$$

which is rather straightforward.

3.4.2 Case Studies for Power Allocation

In this section, two case studies are investigated.

3.4.2.1 Case I

The error propagation factor β is a small value. In this special case, we can omit the impact of imperfect DRS process and R_n becomes,

$$R_n = \log_2 \det \left(\mathbf{I} + (\sigma^2 \mathbf{I})^{-1} \mathbf{H}_n \mathbf{W}_n \mathbf{W}_n^H \mathbf{H}_n^H \right). \tag{3.25}$$

Since the sum rate $(R_n + R_k)$ is a monotone increasing function of P_n, as shown in [171], we only need to find the minimum power for the weak user k to meet the data rate requirement, and then allocate the rest of the power to UE n. In this case, we can get the optimal power by using bisection search algorithm [170, 171].

3.4.2.2 Case II

β is large. In this case, we may discard the ambient (thermal) noise and R_n is only affected by the residual interference from UE k.

$$R_n = \log_2 \det \left(\mathbf{I} + (\beta \mathbf{H}_n \mathbf{W}_k \mathbf{W}_k^H \mathbf{H}_n^H)^{-1} \mathbf{H}_n \mathbf{W}_n \mathbf{W}_n^H \mathbf{H}_n^H \right). \tag{3.26}$$

This can happen when the received SINR for UE k is relatively small, causing a higher error probability. The sum rate in this case is neither an increasing nor decreasing function of P_n, and hence is difficult to track. As we will see later in the simulation section, sum rate is affected by the choice of β. As a preliminary research, we present some results on how the power allocation will affect the sum rate.

3.5 Numerical Results

In this section, we present our simulation results. The total power of the BS is 2 Watts. The number of BS and UE antennas is both equal to 2. The average channel gain for UE n and UE k is 0 and 5 dB, respectively. As for the small β, we choose $\beta = 0.05$, while the large β equals 0.65. $\sigma = 0.5$ in our system. The minimum data rate for UE k is 1 bits/s/Hz. For comparison purposes, we also list the results with precoding as $\mathbf{W}_k \mathbf{W}_k^H = \frac{P_k}{M} \mathbf{I}_M$ and $\mathbf{W}_n \mathbf{W}_n^H = \frac{P_n}{M} \mathbf{I}_M$.

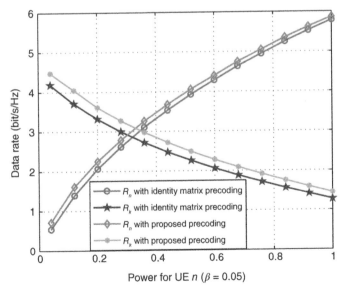

Figure 3.1 UE rate with different precoding matrix as P_n increases ($\beta = 0.05$).

MAXITER = 5 as our iterative precoding algorithm converges very fast. All the results come from 10,000 independent Monte Carlo experiments to ensure the confidence level.

Figure 3.1 shows the rates of UE n and UE k as P_n changes, respectively. β is set to be 0.05 for error propagation in this case. We can see that the rate of UE n increases when P_n increases, while the rate of UE k decreases when P_n increases. It is obvious that the rate of a UE increases when its assigned power increases since the SINR increases. We can also see the rate of UE n increases faster than the rate of UE k when their power increase individually. Since UE n is a user with better channel condition, increasing power slightly can increase the rate a lot. Since β is small, the residual interference from UE k does not affect the performance of UE n too much. UE rates are also shown for identity matrix precoding method. The performance of the identity matrix precoding method has a similar trend to the performance of the proposed precoding design, but the identity matrix precoding method does not perform as well as the proposed precoding design. Another note is the gap between two precoding matrices is small with UE n; this is because as the SINR increases, the water-filling algorithm has a similar performance compared with equal power distribution.

Figure 3.2 shows the sum rate of UE n and UE k as P_n changes when $\beta = 0.05$. We can see that the sum rate increases while P_n increases. From Figure 3.1, it has been shown that the rate of UE n increases faster than the decreasing speed of rate of UE k when P_n increases. Therefore, the sum rate of UE n and UE k increases while P_n increases. Figure 3.2 also shows the proposed precoding design performs better than identity matrix precoding method with respect to sum rate.

Figure 3.3 shows the rates of UE n and UE k as P_n changes when β is set to be 0.65 for error propagation. We can still see that the rate of UE n increases when P_n increases, while the rate of UE k decreases when P_n increases. However, the rate of UE n increases faster than the decreasing speed of rate of UE k when P_n increases. The rate of UE n increases

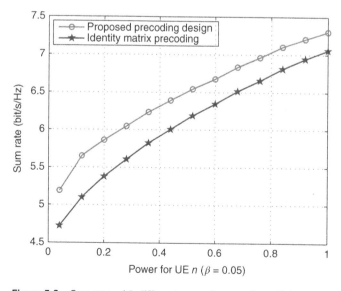

Figure 3.2 Sum rate with different precoding matrix as P_n increases ($\beta = 0.05$).

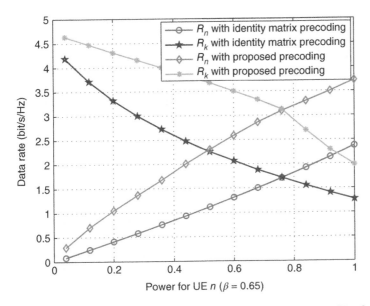

Figure 3.3 UE rate with different precoding matrix as P_n increases ($\beta = 0.65$).

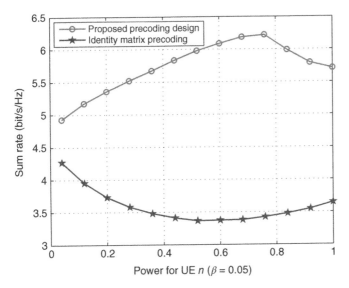

Figure 3.4 Sum rate with different precoding matrix as P_n increases ($\beta = 0.65$).

slower than the decreasing speed of rate of UE k when $P_n > 0.76$. The reason is that when $P_n < 0.76$, the interference from UE n is smaller than the noise; therefore, the SINR decreases slowly. However, when $P_n > 0.76$, the interference becomes dominant and causes the SINR to decrease rapidly. Compared with Figure 3.1, the rate of UE n increases slower than that in Figure 3.1. Since a bigger β is used in this case, residual interference from UE k has a bigger effect to UE n. Therefore, the rate of UE n increases slower because of stronger residual interference from UE k. For the identity matrix precoding method, the rate of UE n increases slower than the rate of UE k when their power increases individually because UE n has a strong residual interference from UE k. We can see that the proposed precoding design performs much better than the identity matrix precoding method because it is designed to optimize the UE sum rate.

Figure 3.4 shows the sum rate of UE n and UE k as P_n changes when $\beta = 0.65$. We can see that the proposed precoding design performs much better than the identity matrix precoding method because it is designed to optimize the UE sum rate.

3.6 Summary

In this chapter, we consider a downlink wireless network which jointly incorporates NOMA and MIMO. A sum rate optimization problem is formulated with error propagation in SIC. In order to solve the problem, we introduce the concept of equivalent channel and propose a sequential solution which solves for precoding matrix by applying an iterative algorithm first. Then we investigate the impact of power allocation by two cases: small error propagation factor and large error propagation factor. Simulations are performed to verify the superiority of proposed precoding design and our analyses on power allocation with residual interference.

4

NOMA in Relay and IoT for 5G Wireless Networks

4.1 Outage Probability Study in a NOMA Relay System

4.1.1 Background

In Chapters 2 and 3, we have shown NOMA in D2D underlaid MIMO networks and the SIC error propagation. Existing works also evaluated NOMA's performance under outage analysis. In [214], the authors analyzed the performance of NOMA theoretically and they concluded that the disparity, either from user channels or intentionally created by allocating different power factor, can further be beneficial to the system performance. A similar conclusion was drawn from cognitive radio NOMA (CR-NOMA) in [45]. Outage probability is a metric widely used in performance evaluation. It is shown in [44] that the outage performance of NOMA is superior to the traditional OMA in a group of randomly deployed users.

This chapter develops a precoding and power allocation strategy to further enhance the system performance in terms of sum rate. Similarly, both [171] and [170] apply NOMA into MIMO scheme. The algorithms in their studies can be applied with or without CSI. In [159] and [215] system-level performance of using NOMA in LTE and heterogeneous networks is evaluated, and the results show promising improvements over existing radio access technologies (RATs). In [141], random beamforming together with intra-beam superposition coding and SIC with BS cooperation is investigated. Relay cooperative communication has been studied in the following studies. Men and Ge [129] uses a single-antenna amplified-and-forward (AF) relay to help the transmission between multi-antenna BS and users. Kim and Lee [96] uses relay to help the transmission to a poor-channel user. Ding *et al.* [46] investigates the system performance under a selection of multiple relays.

As NOMA uses SIC at the receiver to decode multiple user information, the performance of SIC can greatly impact NOMA. Most existing studies assume perfect SIC in NOMA study with a few like [140] considering SIC error due to imperfect channel knowledge. One of our earlier studies [172] investigates the sum rate performance in a MIMO+NOMA system, and it considers error propagation in the SIC process. The idea was inspired by the decoding process in CDMA systems [8]. It assumes there is a residual power from previously decoded signals, and this residual power can arise due to channel estimation error, imperfect constellation mapping, or channel fading. SIC error propagation causes a chain effect and it affects the last decoded user most.

5G and Beyond Wireless Communication Networks, First Edition. Haijian Sun, Rose Qingyang Hu, and Yi Qian.
© 2024 John Wiley & Sons Ltd. Published 2024 by John Wiley & Sons Ltd.

In this chapter, two NOMA relay schemes are presented and evaluated, namely NOMA cooperative scheme and NOMA TDMA scheme. In NOMA cooperative scheme, the completion of one round information transmission consists of two time slots. In the first slot the BS uses NOMA to send the superimposed signal to two relays. Upon receiving the signal, these two relays will decode the signals by using SIC and then form a cooperative communication pair to send the precoded signals to the respective recipients in the second time slot. Dirty paper code (DPC) is used as precoding at relays to eliminate the inter-user interference in the second time slot. As a comparison, NOMA TDMA scheme uses three times slots to complete one round information transmission. The first time slot does the same as in scheme one. After relays decode the message, the first relay sends one signal to user one in the second slot and the second relay send another signal to user 2 in the following slot. Analytical models on outage performance are derived for both schemes in this chapter.

4.1.2 System Model

The study considers a downlink wireless communication system that consists of one access point (AP) and a number of UEs [175]. Each UE can either function as a relay when needed or as a regular UE. The transmit powers of AP and UE are P_s and P_r, respectively. With NOMA, the AP can communicate with two UEs simultaneously. In the case that the channels between AP and these two UEs are poor, two other UEs are selected as relays for multi-hop cooperative transmission. Relays operate in a half-duplex decode-and-forward (DF) mode. The AP and UEs in the system are equipped with a single antenna. For notational simplicity, we denote AP, relay 1, relay 2, UE 1, and UE 2 with subscripts $b, r1, r2, u1$, and $u2$ in the equations, respectively. Furthermore, it is assumed that channels between the AP and two relays are two independent random variables (RVs) following a complex Gaussian distribution with zero mean but different variances, i.e. $h_{b,r1} \sim \mathcal{CN}(0, \sigma^2_{b,r1})$, $h_{b,r2} \sim \mathcal{CN}(0, \sigma^2_{b,r2})$. Without loss of generality, we assume $|h_{b,r1}|^2 > |h_{b,r2}|^2$ and thus $\alpha_s < \beta_s$ is satisfied to provide sufficient decoding capability for NOMA weaker user. On the other hand, channels between relays and UEs can be modeled as independent complex Gaussian RVs with zero mean and unit variance, i.e. $h_{i,j} \sim \mathcal{CN}(0, 1)$. $i = \{r1, r2\}, j = \{u1, u2\}$.

4.1.2.1 NOMA Cooperative Scheme

Each round of NOMA cooperative transmission consists of two time slots. In the first time slot, the AP transmits a composite signal $x_s = \sqrt{\alpha_s P_s} x_1 + \sqrt{\beta_s P_s} x_2$ according to the NOMA principle, where x_1 and x_2 are signals intended for users 1 and 2, respectively; α_s and β_s are the corresponding power allocation factors and satisfy $\alpha_s + \beta_s = 1$. The received signals at two relays are, respectively, expressed as

$$y_{b,r1} = h_{b,r1}(\sqrt{\alpha_s P_s} x_1 + \sqrt{\beta_s P_s} x_2) + n_{b,r1},\qquad(4.1)$$

and

$$y_{b,r2} = h_{b,r2}(\sqrt{\alpha_s P_s} x_1 + \sqrt{\beta_s P_s} x_2) + n_{b,r2}.\qquad(4.2)$$

$n_{b,r1}$ and $n_{b,r2}$ are additive white Gaussian noise (AWGN) and follow $n_{b,i} \sim \mathcal{CN}(0, N_0)$, $i = \{r1, r2\}$. Both relays use SIC to decode the received signals. We first present the analysis by assuming perfect SIC and the results with imperfect SIC will be presented later. For relay 1, x_2 will be decoded first by treating x_1 as interference and the achievable signal-to-interference-noise ratio (SINR) for x_2 is

$$\gamma_{r1,x2} = \frac{\beta_s P_s |h_{b,r1}|^2}{\alpha_s P_s |h_{b,r1}|^2 + N_0}. \tag{4.3}$$

Relay 1 then subtracts x_2 from the composite signal and decodes x_1 with only AWGN. Thus, the achievable SINR becomes

$$\gamma_{r1,x1} = \frac{\alpha_s P_s |h_{b,r1}|^2}{N_0}. \tag{4.4}$$

Similarly, at relay 2, the SINR for x_2 and x_1 can be expressed as

$$\gamma_{r2,x2} = \frac{\beta_s P_s |h_{b,r2}|^2}{\alpha_s P_s |h_{b,r2}|^2 + N_0}, \tag{4.5}$$

and

$$\gamma_{r2,x1} = \frac{\alpha_s P_s |h_{b,r2}|^2}{N_0}, \tag{4.6}$$

respectively.

In the second time slot, relay 1 transmits x_1 to user 1 while relay 2 transmits x_2 to user 2 by using precoded cooperative transmission. The received signals at users 1 and 2 are expressed as

$$y_{u1} = h_{r1,u1}\hat{x}_1 + h_{r2,u1}\hat{x}_2 + n_{u1}, \tag{4.7}$$
$$y_{u2} = h_{r1,u2}\hat{x}_1 + h_{r2,u2}\hat{x}_2 + n_{u2},$$

where AWGN $n_i \sim \mathcal{CN}(0, N_0), i = \{u1, u2\}$. If we re-write the above equation in the matrix format, we can get $\mathbf{y} = \mathbf{H}\hat{\mathbf{x}} + \mathbf{n}$ and $\mathbf{y} = [y_{u1} \quad y_{u2}]^T$. $\hat{\mathbf{x}} = [\hat{x}_1 \quad \hat{x}_2]^T$ is the precoded transmitted signal vector. The precoding mechanism will be discussed later. $\mathbf{n} = [n_{u1} \quad n_{u2}]^T$ and

$$\mathbf{H} = \begin{bmatrix} h_{r1,u1} & h_{r2,u1} \\ h_{r1,u2} & h_{r2,u2} \end{bmatrix}. \tag{4.8}$$

To further minimize inter-user interference, DPC is applied at relays as the precoding scheme. Assume $\mathbf{H}$ is a full-rank matrix and it can be decomposed as $\mathbf{H} = \mathbf{LQ}$, where $\mathbf{L}$ is a 2×2 lower triangular matrix and $\mathbf{Q}$ is a semi-orthogonal matrix, $\mathbf{QQ}^H = \mathbf{I}_2$. Thus, let $\mathbf{W} = \mathbf{Q}^H \mathbf{G}$ and $\mathbf{G}$ is given as

$$\mathbf{G} = \begin{bmatrix} 1 & 0 \\ -\dfrac{l_{2,1}}{l_{2,2}} & 1 \end{bmatrix}, \tag{4.9}$$

where $l_{i,j}$ is the (i,j)-th entry of matrix $\mathbf{L}$.

The received signals at two users can be expressed as

$$\mathbf{y} = \mathbf{H}\hat{\mathbf{x}} + \mathbf{n} = \mathbf{HWx} + \mathbf{n}$$
$$= \begin{bmatrix} l_{1,1} & 0 \\ 0 & l_{2,2} \end{bmatrix} \begin{bmatrix} x_1 \\ x_2 \end{bmatrix} + \begin{bmatrix} n_{u1} \\ n_{u2} \end{bmatrix}. \tag{4.10}$$

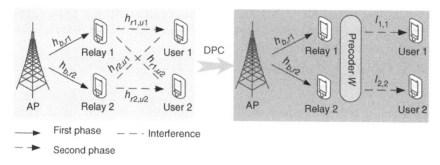

First phase ——▶ · —— · Interference
—— ▶ Second phase

Figure 4.1 NOMA cooperative scheme.

Therefore, SINRs for users 1 and 2 can be written as

$$\gamma_{u1} = \frac{|l_{1,1}|^2 P_r}{N_0}, \gamma_{u2} = \frac{|l_{2,2}|^2 P_r}{N_0}. \tag{4.11}$$

An illustration of NOMA cooperative scheme is shown in Figure 4.1.

For a fair comparison, the user sum rate achieved in one round communication is normalized with respect to the number of time slots in each round. Thus the achievable sum rate for users 1 and 2 is expressed as

$$R_i^{NC} = \frac{1}{2}\log_2(1 + \gamma_i), i = \{u1, u2\}, \tag{4.12}$$

where the factor $1/2$ is used to account for two time slots needed to complete one round transmission.

4.1.2.2 NOMA TDMA Scheme

NOMA TDMA scheme needs three time slots to complete one round communication. The first slot does the same as the first time slot in the NOMA cooperative scheme. Afterwards, relay 1 sends x_1 to user 1 in the second time slot, while relay 2 sends x_2 to user 2 in the third time slot, as shown in Figure 4.2.

Figure 4.2 NOMA TDMA scheme.

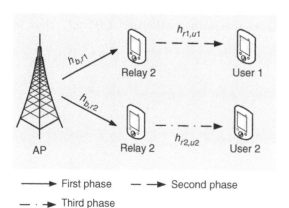

—— ▶ First phase ——▶ Second phase
— · —▶ Third phase

By receiving messages separately in time slots 2 and 3, users 1 and 2 will not experience interference from each other. As a result, the achievable sum rate is

$$R_i^{NT} = \frac{1}{3}\log_2\left(1 + \frac{|h_{j,i}|^2 P_r}{N_0}\right),$$
(4.13)

where $i = \{u1, u2\}, j = \{r1, r2\}$. Likewise, we use the factor $\frac{1}{3}$ to indicate three time slots in this scenario. Note that when calculating the first time slot data rate, we also need to use the factor $1/3$, i.e.

$$R_{i,j} = \frac{1}{3}\log_2(1 + \gamma_{i,j}),$$
(4.14)

where $i = \{r1, r2\}, j = \{x1, x2\}$.

4.1.3 Outage Probability Analysis

In this section, we analyze the system performance in terms of outage probability, which represents the probability of an event that the achieved data rate is less than a predefined one. Outage probability is a good metric for QoS in the system design. A closed-form analytical outage probability is derived for different users, based on which a high SNR approximation will also be presented.

4.1.3.1 Outage Probability in NOMA Cooperative Scheme

Let R_1 and R_2 denote the predefined minimum rates for users 1 and 2, respectively. An outage occurs when the achievable data rate is less than the minimum data rate. Define $\mathcal{O}_{u1,NC}$ as the event of an outage at user 1. We first consider the complementary event of $\mathcal{O}_{u1}^{NC}$, which is denoted as $\mathcal{O}_{u1,NC}^C$. The second time slot transmission relies on the successful decoding at the first time slot. For a DF relaying scheme, $\mathcal{O}_{u1,NC}^C$ happens when relay 1 successfully decodes x_1, and relay 2 successfully decodes x_2, and user 1 successfully decodes x_1. Thus, the outage probability can be calculated as

$$P(\mathcal{O}_{u1,NC}) = 1 - P(\mathcal{O}_{u1,NC}^C)$$
$$= 1 - P\left(\min\{R_{r1,x1}, R_{u1}\} > R_1 \text{ and } \min\{R_{r1,x2}, R_{r2,x2}\} > R_2\right).$$
(4.15)

Similarly, the outage probability for user 2 is

$$P(\mathcal{O}_{u2,NC}) = 1 - P(\mathcal{O}_{u2,NC}^C)$$
$$= 1 - P\left(R_{r1,x1} > R_1 \text{ and } \min\{R_{r1,x2}, R_{r2,x2}, R_{u2}\} > R_2\right).$$
(4.16)

Lemma 4.1 *([68], Theorem 2.3.18) Let $\mathbf{H}$ be a 2×2 matrix and its entries follow i.i.d. Gaussian distribution with zero mean and unit variance. If $\mathbf{H} = \mathbf{LQ}$, where $\mathbf{L}$ is a lower triangle matrix and $\mathbf{Q}$ is a semi-orthogonal matrix, then $|l_{1,1}|^2 \sim \chi^2(4)$ and $|l_{2,2}|^2 \sim \exp(1)$.*

Theorem 4.1 *The outage probabilities for users 1 and 2 in NOMA cooperative scheme can be expressed as*

$$P(\mathcal{O}_{u1,NC}) = 1 - e^{-\frac{\phi_1}{\sigma_{b,r1}^2}} e^{-\frac{\phi_2}{\sigma_{b,r2}^2}} (\phi_3 + 1)e^{-\phi_3},$$
(4.17)

and

$$P(\mathcal{O}_{u2,NC}) = 1 - e^{-\frac{\phi_1}{\sigma_{b,r1}^2}} e^{-\frac{\phi_2}{\sigma_{b,r2}^2}} e^{-\phi_4}, \tag{4.18}$$

where $\rho_s \triangleq \frac{P_s}{N_0}$, $\rho_r \triangleq \frac{P_r}{N_0}$, $z1 \triangleq 2^{2R_1} - 1$, $z2 \triangleq 2^{2R_2} - 1$, $\phi_1 = \max\{\frac{z_1}{\alpha_s \rho_s}, \phi_2\}, \phi_2 = \frac{z_2}{(\beta_s - z_2 \alpha_s)\rho_s}$, $\phi_3 = \frac{z_1}{\rho_r}, \phi_4 = \frac{z_2}{\rho_r}$.

Proof: From equation (4.15),

$$P(\mathcal{O}_{u1,NC}) = 1 - P\left(\min\{R_{r1,x1}, R_{u1}\} > R_1 \text{ and } \min\{R_{r1,x2}, R_{r2,x2}\} > R_2\right)$$

$$= 1 - P\left(\min\left\{\frac{1}{2}\log_2(1 + \gamma_{r1,x1}), \frac{1}{2}\log_2(1 + \gamma_{u1})\right\} > R_1\right) *$$

$$P\left(\min\left\{\frac{1}{2}\log_2(1 + \gamma_{r1,x2}), \frac{1}{2}\log_2(1 + \gamma_{r2,x2})\right\} > R_2\right)$$

$$\overset{a}{=} 1 - P\left(|h_{b,r1}|^2 > \phi_1\right) P\left(|h_{b,r2}|^2 > \phi_2\right) P\left(|l_{1,1}|^2 > \phi_3\right)$$

$$\overset{b}{=} 1 - e^{-\frac{\phi_1}{\sigma_{b,r1}^2}} e^{-\frac{\phi_2}{\sigma_{b,r2}^2}} (\phi_3 + 1)e^{-\phi_3}.$$

Here, $\overset{a}{=}$ holds when $\beta_s > \max\{z_2\alpha_s, \alpha_s\}$. $\overset{b}{=}$ holds since both $|h_{b,r1}|^2$ and $|h_{b,r2}|^2$ follow an exponential distribution with parameter 1 while $|l_{1,1}|^2$ follows a chi-squared distribution with a degree of freedom 4.

Similarly,

$$P(\mathcal{O}_{u2,NC}) = 1 - P\left(R_{r1,x1} > R_1 \text{ and } \min\{R_{r1,x2}, R_{r2,x2}, R_{u2}\} > R_2\right)$$

$$= 1 - P\left(\frac{1}{2}\log_2(1 + \gamma_{r1,x1}) > R_1\right) *$$

$$P\left(\min\left\{\frac{1}{2}\log_2(1 + \gamma_{r1,x2}), \frac{1}{2}\log_2(1 + \gamma_{r2,x2}), \frac{1}{2}\log_2(1 + \gamma_{u2})\right\} > R_2\right)$$

$$= 1 - P\left(|h_{b,r1}|^2 > \phi_1\right) P\left(|h_{b,r2}|^2 > \phi_2\right) P\left(|l_{2,2}|^2 > \phi_4\right)$$

$$= 1 - e^{-\frac{\phi_1}{\sigma_{b,r1}^2}} e^{-\frac{\phi_2}{\sigma_{b,r2}^2}} e^{-\phi_4}.$$

Since $\lim_{x \to 0}(1 - e^{-x}) \simeq x$, in the high SNR regime, i.e. when $\rho_s, \rho_r \to \infty$, user 2 outage probability at high SNR can be approximated as:

$$P(\mathcal{O}_{u2,NC}) = \frac{\phi_1}{\sigma_{b,r1}^2} + \frac{\phi_2}{\sigma_{b,r2}^2} + \phi_4. \tag{4.19}$$

4.1.4 Outage Probability in NOMA TDMA Scheme

As previously stated the first time slot in this scheme also uses NOMA transmission from the AP to two relays. Afterwards two relays will transmit x_1 and x_2 to the respective recipient in the following two time slots separately. Similar to NOMA cooperative scheme, the expressions for outage probabilities for users 1 and 2 are, respectively, expressed as

$$P(\mathcal{O}_{u1,NT}) = 1 - P\left(\min\{R_{r1,x1}, R_{u1}\} > R_1 \text{ and } R_{r1,x2} > R_2\right), \tag{4.20}$$

and

$$P(\mathcal{O}_{u2,NT}) = 1 - P\left(\min\{R_{r2,x2}, R_{u2}\} > R_2\right). \tag{4.21}$$

We have the following theorem for the outage probabilities.

Theorem 4.2 *The outage probabilities for users 1 and 2 in NOMA TDMA scheme can be calculated as*

$$P(\mathcal{O}_{u1,NT}) = 1 - e^{-\frac{\phi_5}{\sigma_{b,r1}^2}} e^{-\phi_6}, \tag{4.22}$$

and

$$P(\mathcal{O}_{u2,NT}) = 1 - e^{-\frac{\phi_7}{\sigma_{b,r2}^2}} e^{-\phi_8}, \tag{4.23}$$

where $\phi_5 = \max\{\frac{z_3}{\alpha_s \rho_s}, \phi_7\}$, $\phi_6 = \frac{z_3}{\rho_r}$, $\phi_7 = \frac{z_4}{(\beta_s - z_4\alpha_s)\rho_s}$, $\phi_8 = \frac{z_4}{\rho_r}$, $z_3 = 2^{3R_1} - 1$, *and* $z_4 = 2^{3R_2} - 1$.

The proof is similar to **Theorem 4.1** and thus is not detailed here. Note that in order for this theorem to hold, we need to have $\beta_s > \max\{z_4\alpha_s, \alpha_s\}$.

4.1.5 Outage Probability with Error Propagation in SIC

In Section 4.1.3, we have derived the outage performance for users 1 and 2 by assuming both relays can decode NOMA signals correctly by using SIC. In what follows, we introduce the concept of error propagation in SIC, which can affect the system performance such as sum rate and outage probability.

The process of SIC consists of decoding, reconstruction, and subtraction (DRS) [172]. Take relay 1 as an example; upon receiving the superimposed signal, x_2 will be decoded first by treating x_1 as interference. Then a reconstruction process will take place where relay 1 estimates its channel gain and uses the decoded signal $\hat{x}_2$. Therefore, the superposition signal for the next decoding symbol x_1 will be updated to

$$y_{r1,x1} = y_{b,r1} - \hat{h}_{b,r1}\hat{x}_2, \tag{4.24}$$

where $\hat{h}_{b,r1}$ is the estimated channel gain for relay 1. Existing studies assume the perfect decoding and cancellation of x_2, and thus we have $y_{r1,x1} = h_{b,r1}\sqrt{\alpha_s P_s}x_1 + n_{b,r1}$. We argue that this is a strong assumption since neither the channel estimation nor signal decoding can be perfect. While we desire to let $\hat{h}_k$ and $\hat{s}_M$ as close to h_k and s_M as possible, factors such as synchronization, phase ambiguity, and deep fading can seriously degrade the SIC process and errors can be accumulated and affect the UE to be decoded afterwards. We refer this process as EP.

θ is defined as the EP factor in this chapter. Since there is a residual power when decoding the second signal, (4.4) and (4.6) can be updated to,

$$\gamma_{r1,x1}^{EP} = \frac{\alpha_s P_s|h_{b,r1}|^2}{N_0 + \theta\beta_s P_s|h_{b,r1}|^2}, \tag{4.25}$$

$$\gamma_{r2,x1}^{EP} = \frac{\alpha_s P_s|h_{b,r2}|^2}{N_0 + \theta\beta_s P_s|h_{b,r2}|^2}.$$

θ represents the amount of residual power from the previous decoding and $0 \le \theta \le 1$. When $\theta = 0$, the results agree with perfect cancellation. $\theta = 1$ is the worst case when SIC fails to decode the first signal and the second stage decoding has to treat the entire first signal as interference. Besides, θ should be inversely proportional to the SNR of x_2. In this chapter, we assume θ is a constant for simplicity.

Similarly, the outage probability analysis is given as follows.

4.1.5.1 Outage Probability in NOMA Cooperative Scheme with EP

Define $\mathcal{O}_{i,NC}^{EP}, i = \{u1, u2\}$ as the outage event of user i in the NOMA cooperative scheme. Then, we have the following theorem for the outage probability.

Theorem 4.3 *The outage probabilities for users 1 and 2 in the NOMA cooperative scheme when considering EP in SIC are, respectively, derived as*

$$P(\mathcal{O}_{u1,NC}^{EP}) = 1 - e^{-\max\{\frac{z1}{(\alpha_s - z_1 \theta \beta_s)\rho_s}, \phi_2\}/\sigma_{b,r1}^2} e^{-\phi_2/\sigma_{b,r2}^2}(\phi_3 + 1)e^{-\phi_3}, \qquad (4.26)$$

and

$$P(\mathcal{O}_{u2,NC}^{EP}) = 1 - e^{-\max\{\frac{z1}{(\alpha_s - z_1 \theta \beta_s)\rho_s}, \phi_2\}/\sigma_{b,r1}^2} e^{-\phi_2/\sigma_{b,r2}^2} e^{-\phi_4}. \qquad (4.27)$$

4.1.5.2 Outage Probability in NOMA TDMA Scheme with EP

Similarly, we also consider EP in SIC in the second scheme. Let $\mathcal{O}_{i,NT}^{EP}, i = \{u1, u2\}$ be the event of an outage. We have the following theorem for the analytical results of outage probabilities.

Theorem 4.4 *The outage probabilities for users 1 and 2 in NOMA TDMA transmission when considering EP in SIC are*

$$P(\mathcal{O}_{u1,NT}^{EP}) = 1 - e^{-\max\{\frac{z3}{(\alpha_s - z_3 \theta \beta_s)\rho_s}, \phi_7\}/\sigma_{b,r1}^2} e^{-\phi_6}, \qquad (4.28)$$

and

$$P(\mathcal{O}_{u2,NT}^{EP}) = 1 - e^{-\frac{\phi_7}{\sigma_{b,r2}^2}} e^{-\phi_8}. \qquad (4.29)$$

Remark 4.1 The constraint for **Theorem 4.3** to hold is $\frac{\alpha_s}{z_1\theta} > \beta_s > \max\{z_2\alpha_s, \alpha_s\}$, which has one additional constraint $(\alpha_s > z_1\theta\beta_s)$ compared with **Theorem 4.1**. Similarly, **Theorem 4.4** holds when $\frac{\alpha_s}{z_3\theta} > \beta_s > \max\{z_4\alpha_s, \alpha_s\}$, which also posts another constraint. These additional constraints can potentially increase the outage probability.

Remark 4.2 We show that power allocation factors impact the outage probability. Specifically, if the constraints in *Remark 4.1* cannot be satisfied, the outage probability will always be 1 for both users, which indicates the failure of both schemes. The reason is that if x_2 cannot be decoded in the first time slot, then the second or third time slot cannot proceed. If the maximum value of β_s is less than $\frac{\alpha_s}{z_1}\Omega_1$, where $\Omega_1 = \min\{z_2(1 + z_1), \frac{1}{\theta}, \frac{z_2(1+z_1)}{1+z_2\theta}\}$, the outage probability under NOMA cooperative scheme is the same for both with EP or without EP cases. Likewise, for NOMA TDMA scheme, the outage probability is the same for user 1 in both EP or no EP cases when $\beta_s < \frac{\alpha_s}{z_3}\Omega_2$, where $\Omega_2 = \min\{z_4(1 + z_3), \frac{1}{\theta}, \frac{z_4(1+z_3)}{1+z_4\theta}\}$. The reason is that if we limit the value of β_s, the bottleneck of the data rate does not come from the first time slot transmission, which may be directly affected by the EP. However, for user 2, the outage probability is always identical with or with not EP since under this circumstance user 2 is not impacted by EP.

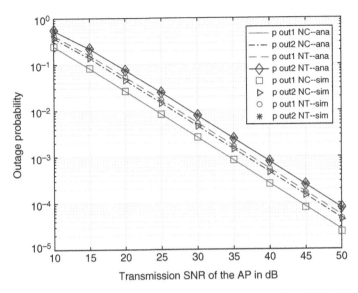

Figure 4.3 Theorem 4.1 and 4.2. $\alpha_s = 0.2, \beta_s = 0.8. R_1 = R_2 = 0.5$ bps/Hz.

4.1.6 Numerical Results

In this section, performance evaluation on the proposed schemes are provided based on both simulation and analysis. Some basic parameters are set as follows. The channel gains of $h_{b,r1}$ and $h_{b,r2}$ are 5 and 1, respectively, i.e. $\sigma^2_{b,r1} = 5, \sigma^2_{b,r2} = 1$. The transmission SNR of the AP ranges from 10 to 50 dB, and the transmit power of both relays is set to half of the AP's power, which means there is a 3 dB difference between P_s and P_r.

Figure 4.3 illustrates the outage performance in both schemes with perfect SIC, i.e. no EP in SIC, as a function of the AP transmit SNR in dB. The predefined minimum data rates R_1 and R_2 are both set to 0.5 bps/Hz. Besides, $\alpha_s = 0.2$ and $\beta_s = 0.8$ are constraints. Apparently, optimizing α_s and β_s based on channel condition and transmit SNR will further improve the outage probability performance, and this can be explored in the future work. It is observed that all the outage probabilities decrease with the increment of SNR. The analytical results match the simulation results very well, which validates the earlier analysis in **Theorem 4.1** and **Theorem 4.2**. Because of this, we only present the analytical results for better illustrations in the figures for the following parts.

Further, by comparing the performance of NOMA cooperative and NOMA TDMA schemes, one can conclude that NOMA cooperative scheme achieves lower outage probabilities than the NOMA TDMA scheme, which uses three time slots in one round communication, and hence the added factor $\frac{1}{3}$ decreases the total sum rate. In both schemes, user 1 outperforms user 2 since user 2's message x_2 is decoded first, which has a higher interference term.

Figure 4.4 presents the result for **Theorem 4.3**. A new set of parameters is selected to satisfy *Remark 4.1* and *Remark 4.2*. The corresponding parameters are $\alpha_s = 0.36$,

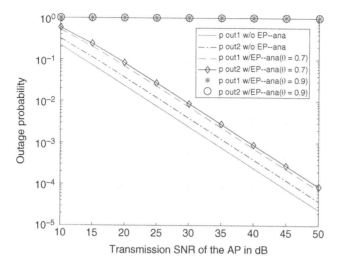

Figure 4.4 Theorem 4.3. $\alpha_s = 0.36, \beta_s = 0.64$. $R_1 = R_2 = 0.4$ bps/Hz. $\theta = 0.7$ and $\theta = 0.9$.

$\beta_s = 0.64, R_1 = R_2 = 0.4$ bps/Hz. The curve without EP is also plotted for reference. One can see that SIC EP degrades the outage performance largely when $\theta = 0.7$. However, when $\theta = 0.9$, the condition $\frac{\alpha_s}{z_1\theta} > \beta_s$ is not satisfied any more, making both the analytical and simulated outage probabilities to 1.

The result for **Theorem 4.4** is shown in Figure 4.5. Likewise, we plot the case without EP for reference. The parameters for this scheme are $\alpha_s = 0.36, \beta_s = 0.64, R_1 = R_2 = 0.4$ bps/Hz. These parameters are selected to meet the requirements of *Remark 4.1* and *Remark 4.2*. When EP is considered, the performance becomes worse for user 1, while user 2

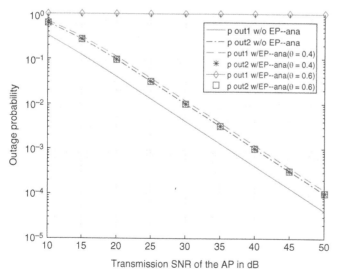

Figure 4.5 Theorem 4.4. $\alpha_s = 0.36, \beta_s = 0.64$. $R_1 = R_2 = 0.4$ bps/Hz. $\theta = 0.4$ and $\theta = 0.6$.

outage probability remains the same. When $\theta = 0.6$, the condition $\frac{\alpha_s}{z_1 \theta} > \beta_s$ is not satisfied. As expected, the outage probability becomes 1 for user 1.

4.2 NOMA in a mmWave-Based IoT Wireless System with SWIPT

4.2.1 Introduction

The unprecedented growth of mobile devices including smart phones, tablets, laptops, and IoT devices drives the wireless telecommunication industry to a new level. The requirements come from various aspects such as higher data rate, fairness, tremendous connectivity, and low latency from different applications and various end users. Therefore, as a new generation technology, 5G emerges with its goal to provide 1000 times higher data rate, 1 ms low latency, and support billions of upcoming IoT devices. Among these features, 1000 times capacity can be achieved by the new mmWave spectrum, novel network architectures, and new radio access technologies (RATs) [78].

Due to the ad hoc deployment nature of most low-power nodes and devices, they may have limited access to wireline power charging facilities and also have limited battery life. In this section, low-power relay nodes and devices are assumed to be capable of energy harvest functionality. More specifically, simultaneous wireless information and power transfer (SWIPT) is considered. SWIPT can have two implementation modes, namely time switching (TS) mode and power splitting (PS) mode [108]. In the TS mode, a dedicated resource is used for energy transfer from which the harvested energy is then used for future information transmission. In the PS mode, upon receiving the radio signal, the energy harvest node splits the signal into two parts. The first part is used for signal decoding, while the second part is used for energy charging. A linear energy harvest model, which assumes the output power of the energy harvest circuit grows linearly with the input power, is applied in most existing works. Cooperative NOMA system with SWIPT is studied in [110], where they proposed different user selection schemes and evaluated the performance with outage probability. This paradigm is proved impractical based on field test results as shown in [21]. As a result, a more practical yet more complicated non-linear model which better matches current circuit design is considered in this chapter. Thus the wireless heterogeneous system in this study consists of higher-power MBSs and low-power relays with SWIPT that is based on the non-linear energy harvesting model. Downlink NOMA is first used to transmit composite signals to UE and relay. Relay then harvests the energy by using non-linear model in PS mode. With the harvested energy, relay sends the received signal to the cell edge UE.

4.2.2 System Model

The system model is based on a mmWave downlink wireless heterogeneous system that consists of high-power MBSs, low-power relays, and low-power IoT devices, such as sensors or wearable devices [176]. At mmwave band, MBSs are equipped with a large number of antennas, which have narrow half-power-beamwidth (HPBW) to combat with the severe pathloss and each transmission is conducted with a single antenna. While each low-power relay or IoT device is equipped with a single antenna due to the size and power constraints. It is assumed that MBSs can coordinate the transmission direction; hence, inter-cell and intra cell interference can be eliminated by carefully aligning the beam directions. Furthermore, relaying and NOMA are used to help reach UEs out of coverage due to severe blockage at mmWave band. Without loss of generality, IoT UE 1 and IoT UE 2 are selected, where UE 1 is in the beamforming coverage area while there is a severe blockage between BS and UE 2 so that a direct transmission link between the MBS and UE 2 is difficult to establish. Thus BS can communicate to UE 2 through relays. In this chapter we assume D2D relaying mode is used so that the relay can communication with a UE in close proximity and we assume the relay is capable of rechargeable functionality. So the power consumed for relaying comes directly from electromagnetic waves, which can relieve the concern on limited battery life for typical IoT devices. With NOMA and relay, complete transmission cycle consists of two phases. In the first phase, the BS sends a composite signal to UE 1 and a selected relay device simultaneously by applying NOMA. After receiving the signal, the relay device splits the signal into two parts. One part is for information decoding and the other part is for energy harvesting. In the second phase, the BS sends another message to UE 1 while the relay device sends the decoded message to UE 2 by using the harvested energy in phase 1.

Denote the channel between BS and UE 1, BS and relay device, relay device and UE 2, as h'_{B1}, h'_{BR}, and h'_{R2}, respectively. Frequency flat quasi-static block fading model is used here, so the channel does not change during the two transmission phases while the channel changes from cycle to cycle. Additionally, $h'_i = \dfrac{h_i \sqrt{a_0}}{\sqrt{1+d_i^\alpha}}$, where h_i is modeled as Rayleigh fading with $h_i \sim \mathcal{CN}(0,1), i = \{B1, BR, R2\}$ [47]. a_0 is antenna-specific gain for the BS and $a_0 = 1$ when $i = R2$. An illustration of the system model is in Figure 4.6. In the following, the transmission process for each cycle is illustrated.

4.2.2.1 Phase 1 Transmission
In this phase, the BS sends the superimposed message to both UE 1 and the relay. The message is given as $x = \sqrt{\lambda_1 P_{BS}} x_1 + \sqrt{\lambda_2 P_{BS}} x_2$, where λ_1 and λ_2 are power allocation factors for UE 1 and the relay, respectively, with $\lambda_1 + \lambda_2 = 1$. x_1 and x_2 are normalized intended

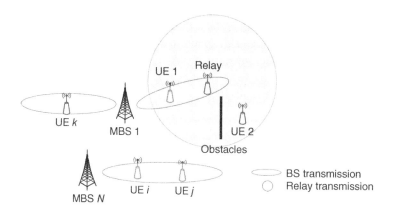

Figure 4.6 System model.

signal for UE 1 and UE 2. P_{BS} is the transmission power of the BS. At the receiver side, UE 1 observes y^1_{UE1}, which is expressed as

$$y^1_{UE1} = \frac{h_{B1}\sqrt{a_0}}{\sqrt{1+d^\alpha_{B1}}}x + n_{B1}$$

$$= \frac{h_{B1}\sqrt{a_0}}{\sqrt{1+d^\alpha_{B1}}}(\sqrt{\lambda_1 P_{BS}}x_1 + \sqrt{\lambda_2 P_{BS}}x_2) + n_{B1}, \qquad (4.30)$$

where n_{B1} is the additive Gaussian white noise (AWGN) with variance σ^2, d_{B1} is the distance from the BS to UE 1, α is the path loss exponent for line-of-sight (LOS).

Without loss of generality, we assume $|h_{B1}|^2 > |h_{BR}|^2$. Hence according to NOMA protocol, $\lambda_1 < \lambda_2$ is set to ensure QoS at the weak receiver. With this setting, UE 1 first decodes signal x_2 with its SINR formulated as

$$\gamma^1_{UE1,x_2} = \frac{\lambda_2 \rho_{B1}|h_{B1}|^2}{\lambda_1 \rho_{B1}|h_{B1}|^2 + 1}, \qquad (4.31)$$

where $\rho_{B1} = \frac{P_{BS}a_0}{\sigma^2(1+d^\alpha_{B1})}$ is the transmission SNR from the BS to UE 1. The superscript "1" indicates the first phase. SIC is performed to remove x_2 from the superimposed signal, then UE 1 can decode its own message with the following SINR

$$\gamma^1_{UE1,x_1} = \lambda_1 \rho_{B1}|h_{B1}|^2. \qquad (4.32)$$

At the relay side, it first splits the observation into two parts. One part is for the rechargeable unit, which consists of a super capacitor or a short-term high efficiency battery. The other part is for information decoding, which can be expressed as

$$y^D_R = \frac{h_{BR}\sqrt{a_0}}{\sqrt{1+d^\alpha_{BR}}}x\sqrt{1-\beta} + n_{BR}$$

$$= \frac{h_{BR}\sqrt{a_0}}{\sqrt{1+d^\alpha_{BR}}}\sqrt{1-\beta}(\sqrt{\lambda_1 P_{BS}}x_1 + \sqrt{\lambda_2 P_{BS}}x_2) + n_{BR}, \qquad (4.33)$$

where β is the power split coefficient indicating the portion of power assigned to energy harvest unit. n_{BR} has the same distribution with n_{B1}. Signal y^D_R goes through the decoding unit for x_2, the corresponding SINR is

$$\gamma^1_{R,x_2} = \frac{(1-\beta)\lambda_2 \rho_{BR}|h_{BR}|^2}{(1-\beta)\lambda_1 \rho_{BR}|h_{BR}|^2 + 1}, \qquad (4.34)$$

where $\rho_{BR} = \frac{P_{BS}a_0}{\sigma^2(1+d^\alpha_{BR})}$ is the transmission SNR from the BS to the relay.

The remaining power $P^C_R = |h_{BR}|^2\beta\rho_{BR}\sigma^2$ is harvested by the relay. In this chapter, we adopt the non-linear energy harvest model, which is more precise in modeling the power-in-power-out relation in current wireless charging technology. Specifically, the harvested energy can be expressed as a logistic (sigmoidal) function

$$P^{EH}_R = \frac{M}{1+\exp\left(-a(P^C_R - b)\right)}, \qquad (4.35)$$

where M, a, b are constants and represent different physical meanings in wireless charging. M denotes the maximum harvested power at the relay when the energy harvesting circuit

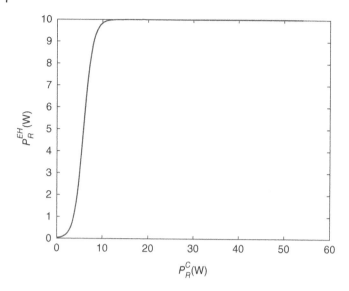

Figure 4.7 Power-in-power-out response in the non-linear energy harvest model.

is saturated. a together with b capture the joint effect of resistance, capacitance, and circuit sensitivity [19].

Boshkovska *et al.* [21] provides a more sophisticated model, which captures the zero-input-zero-output feature in wireless charging and can be modeled in the following.

$$P_R^{EH} = \frac{\Psi - M\Omega}{1 - \Omega}, \quad \Omega = \frac{1}{1 + \exp(ab)}, \tag{4.36}$$

where $\Psi = \frac{M}{1 + \exp(-a(P_R^C - b))}$.

In the subsequent analysis, we use model (4.35) based on the following reasons. (i) Our model does not have zero-power input case; (ii) The general logistic function can reduce the complexity in outage analysis; (iii) (4.35) can provide sufficient precision.

Figure 4.7 presents the power-in-power-out relation with 1000 independent events, based on which the parameters are estimated as follows, $\beta = 0.6$, $\sigma = 0.0995$, $M = 10$, $a = 1$, $b = \beta \rho_{BR}\sigma^2$, and $\rho_{BR} = 30$ dB.

4.2.2.2 Phase 2 Transmission

During the second phase, the relay sends x_2 to UE 2 with the energy harvested in Phase 1. Meanwhile, the BS sends another signal x_3 to UE 1. The received signal at UE 1 and UE 2 is expressed as

$$y_{UE1}^2 = \sqrt{P_{BS}}\frac{h_{B1}\sqrt{a_0}}{\sqrt{1 + d_{B1}^\alpha}}x_3 + \sqrt{P_R^{EH}}\frac{h_{R1}}{\sqrt{1 + d_{R1}^\alpha}}x_2 + n_{B1}, \tag{4.37}$$

and

$$y_{UE2}^2 = \sqrt{P_R^{EH}}\frac{h_{R2}}{\sqrt{1 + d_{R2}^\alpha}}x_2 + n_{B2}, \tag{4.38}$$

respectively.

Since UE 1 already decodes x_2 in Phase 1, by appropriately estimating the channel h_{R1}, it can employ SIC to subtract x_2 from its observation [96]. The remaining SINR becomes

$$\gamma^2_{UE1,x_3} = \rho_{B1}|h_{B1}|^2. \tag{4.39}$$

For UE 2, since there is a severe blockage between BS and itself, it has a negligible interference from BS. The achievable SINR at UE 2 is

$$\gamma^2_{UE2,x_2} = \rho_{EH}|h_{R2}|^2, \tag{4.40}$$

with $\rho_{EH} = \dfrac{P_R^{EH}}{\sigma^2(1+d_{R2}^\alpha)}$.

4.2.3 Outage Analysis

In this section, we will provide mathematical analysis on the outage probability of the proposed scheme. The outage probability is defined as the probability of events where certain measurements such as SINR or data rate cannot meet the pre-defined threshold.

4.2.3.1 UE 1 Outage Probability

Define the minimum data rates for messages x_1, x_2, and x_3 as R_1, R_2, and R_3, respectively. Below the minimum data rate, a UE will have an outage. Since UE 1 involves in both phases, outage occurs when UE 1 fails to decode x_2 and x_1 in phase 1 or fails to decode x_3 in phase 2. For simplicity, we can consider the complementary event first. Specifically, we can derive the outage probability of UE 1 as follows.

$$P(\mathcal{O}_{UE1}) = 1 - P(\mathcal{O}^C_{UE1})$$
$$= 1 - P\left(\frac{1}{2}\log_2(1 + \gamma^1_{UE1,x_2}) > R_2 \text{ and } \frac{1}{2}\log_2(1 + \gamma^1_{UE1,x_1}) > R_1\right.$$
$$\left. \text{and } \frac{1}{2}\log_2(1 + \gamma^2_{UE1,x_3}) > R_3\right).$$

Notice that channel $h_{B1} \sim \mathcal{CN}(0, 1)$ and $|h_{B1}|^2 \sim \exp(1)$. Define $z_1 = 2^{2R_1} - 1, z_2 = 2^{2R_2} - 1$ and $z_3 = 2^{2R_3} - 1$.

$$P(\mathcal{O}_{UE1}) = P(|h_{B1}|^2 > \phi_1)$$
$$= 1 - e^{-\phi_1}, \tag{4.41}$$

where $\phi_1 = \max\left\{\dfrac{z_2}{\lambda_2\rho_{B1}-z_2\lambda_1\rho_{B1}}, \dfrac{z_1}{\lambda_1\rho_{B1}}, \dfrac{z_3}{\rho_{B1}}\right\}$.

Note that the above outage probability is conditioned on $\lambda_2 > z_2\lambda_1$. Otherwise the outage occurs with probability 1.

4.2.3.2 UE 2 Outage Probability

For UE 2, since the BS only transmits x_2 via the relay. Thus the bottleneck of this transmission depends on the minimum data rate in two phases. The outage probability for UE 2 is

$$P(\mathcal{O}_{UE2}) = 1 - P(\mathcal{O}^C_{UE2})$$
$$= 1 - P\left(\min\left\{\frac{1}{2}\log(1 + \gamma^1_{R,x_2}), \frac{1}{2}\log(1 + \gamma^2_{UE2,x_2})\right\} > R_2\right)$$
$$= 1 - P\left(\min\{\gamma^1_{R,x_2}, \gamma^2_{UE2,x_2}\} > z_2\right). \tag{4.42}$$

The following theorem provides an analytical result for the outage probability of UE 2.

Theorem 4.5 *The outage probability for UE 2 in the proposed non-linear energy harvest model is $P(\mathcal{O}_{UE2}) = 1 - \frac{c_2}{c_4}e^{-c_1}(c_3e^{-c_1c_4})^{-\frac{1}{c_4}}\Gamma(\frac{1}{c_4}, c_3e^{-c_1c_4})$, where c_1, c_2, c_3 and c_4 are constants and defined in the following proof.*

Proof: Let $c = (1 + d_{R2}^\alpha)$, the outage probability becomes

$$P(\mathcal{O}_{UE2}) = 1 - P(\min\{\gamma_{R,x_2}^1, \gamma_{UE2,x_2}^2\} > z_2)$$
$$= 1 - P(\gamma_{R,x_2}^1 > z_2, \gamma_{UE2,x_2}^2 > z_2). \tag{4.43}$$

Let probability $P(\gamma_{R,x_2}^1 > z_2, \gamma_{UE2,x_2}^2 > z_2)$ be P_1 for conciseness. Furthermore, let $|h_{BR}|^2 = x$ and $|h_{R2}|^2 = y$. x and y both follow an exponential distribution with parameter 1, and they are independent to each other.

$$P_1 = P\left(\frac{(1-\beta)\lambda_2\rho_{BR}x}{(1-\beta)\lambda_1\rho_{BR}x + 1} > z_2, \frac{P_R^{EH}}{\sigma^2 c}y > z_2\right)$$

$$\overset{a}{=} P\left(x > \frac{z_2}{(1-\beta)\rho_{BR}(\lambda_2 - \lambda_1 z_2)}, \right. \tag{4.44}$$

$$\left. \frac{M}{\sigma^2 c(1 + \exp(-a(\beta\rho_{BR}\sigma^2 x - b)))}y > z_2\right),$$

where $\overset{a}{=}$ is conditioned on $\lambda_2 > \lambda_1 z_2$. Otherwise the outage probability will be always equal to one, as already observed in the existing literature. Define $f(x) = \frac{M}{\sigma^2 c(1 + \exp(-a(\beta\rho_{BR}\sigma^2 x - b)))}$ and let $c_1 = \frac{z_2}{(1-\beta)\rho_{BR}(\lambda_2 - \lambda_1 z_2)}$. The above joint probability can be evaluated as

$$P_1 = \int_{c_1}^\infty \int_{\frac{z_2}{f(x)}}^\infty e^{-x}e^{-y}dxdy$$

$$= \int_{c_1}^\infty \exp\left(-x - \frac{z_2}{f(x)}\right)dx. \tag{4.45}$$

$$= e^{-\frac{z_2\sigma^2 c}{M}}\int_{c_1}^\infty \exp\left(-x - \frac{z_2\sigma^2 c}{M}e^{ab}\exp(-a\beta\rho_{BR}\sigma^2 x)\right)dx.$$

For notation simplicity, define $c_2 = e^{-\frac{z_2\sigma^2 c}{M}}$, $c_3 = \frac{z_2\sigma^2 c}{M}e^{ab}$ and $c_4 = a\beta\rho_{BR}\sigma^2$. Then P_1 can be simplified as

$$P_1 = c_2\int_{c_1}^\infty \exp(-c_3e^{-c_4x} - x)dx. \tag{4.46}$$

Let $u = xc_4 - c_1c_4, u \in [0, \infty]$. According to ([65], 3.331-1)

$$P_1 = \frac{c_2}{c_4}e^{-c_1}\int_0^\infty \exp\left(-c_3e^{-c_1c_4}e^{-u} - \frac{u}{c_4}\right)du$$

$$= \frac{c_2}{c_4}e^{-c_1}(c_3e^{-c_1c_4})^{-\frac{1}{c_4}}\Gamma\left(\frac{1}{c_4}, c_3e^{-c_1c_4}\right). \tag{4.47}$$

$\Gamma(\mu_2, \mu_1)$ is the lower incomplete gamma function, which is

$$\Gamma(\mu_2, \mu_1) = \int_0^{\mu_1} e^{-t}t^{\mu_2-1}dt, \tag{4.48}$$

where $\mu_2 > 0$.

4.2.3.3 Outage at High SNR

In this section, we provide the approximation for the outage probability at high SNR region. Specifically, if $\rho_{B1}, \rho_{BR} \to \infty$, the outage probability for UE 1 becomes

$$P(\mathcal{O}_{UE1}^H) = \phi_1 = \max\left\{ \frac{z_2}{\lambda_2\rho_{B1} - z_2\lambda_1\rho_{B1}}, \frac{z_1}{\lambda_1\rho_{B1}}, \frac{z_3}{\rho_{B1}} \right\}, \tag{4.49}$$

since $\lim_{x\to 0}(1 - e^{-x}) \simeq x$.

For UE 2, the maximum charging power is M even when P_R^C becomes infinity. Thus, the high approximation becomes

$$P(\mathcal{O}_{UE2}^H) = 1 - P\left(\frac{\lambda_2}{\lambda_1} > z_2, \frac{M}{\sigma^2(1 + d_{R_2}^\alpha)}|h_{R2}|^2 > z_2 \right). \tag{4.50}$$

When $\frac{\lambda_2}{\lambda_1} > z_2$, the result becomes

$$P(\mathcal{O}_{UE2}^H) = 1 - e^{-\frac{z_2\sigma^2(1+d_{R_2}^\alpha)}{M}}. \tag{4.51}$$

Otherwise, if $\frac{\lambda_2}{\lambda_1} < z_2$, the outage probability will be always one in the high SNR regime.

4.2.3.4 Diversity Analysis for UE 2

Based on the definition of diversity, we have

$$d_{UE2} = -\lim_{\rho_{BR}\to\infty} \frac{\log P(\mathcal{O}_{UE2})}{\log \rho_{BR}} = 0. \tag{4.52}$$

This means in the non-linear energy harvest model, no diversity will be achieved. The reason is that as the input power increases, the power harvested becomes saturated, which limits the further data rate growth, hence the outage probability performance.

4.2.4 Numerical Results

In this section, numerical performance results are presented based on both simulations and analysis. The parameters for evaluation are chosen in the following. $a_0 = 4$, which indicates the horn antenna gain is 6 dB. $\lambda_1 = 0.4, \lambda_2 = 0.6$. $M = 4$, which means the maximum charging power for the relay is 4 Watts. For illustration purposes, the distance d_{BR}, d_{R2}, and d_{B1} are small, which are set to 8, 2, and 10, respectively. Similar settings can also be found in [108]. Furthermore, the predefined thresholds for data rates are $R1 = R3 = 0.5$ bps/Hz and $R2 = 0.3$ bps/Hz.

Figure 4.8 shows the outage probability of UE 1 and UE 2 with regards of the transmission SNR in dB. "ana" stands for analytical result while "sim" is the simulation one. The performance can be optimized by carefully choosing λ_1 and λ_2. The detailed study on how to select λ_1 and λ_2 values to achieve optimal performance is not the focus of this chapter and hence not extended. Further, since a and b can also impact the system performance, the outage probability of UE 2 is evaluated with different a, b values. By fixing $\beta = 0.8$, both the simulation and analytical results are presented. As we can see from Figure 4.8, the analytical results match well with the simulation ones for UE 1. As expected, the outage probability decreases linearly in log scale with the increase of transmission SNR. For UE 2, when $a = 2.5, b = 3$, the outage probability of UE 2 is lower than the case with $a = 6.5, b = 4$, which indicates that energy harvest circuit will affect the system performance. Also, as the

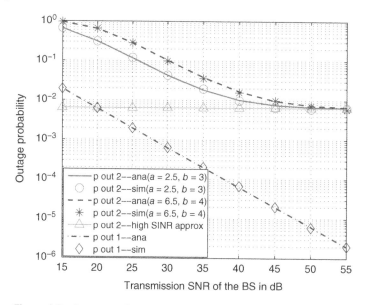

Figure 4.8 Outage performance for both UEs with comparison to analytical results.

transmission SNR becomes larger, the gap becomes less apparent. The reason is as SNR becomes larger, the harvested energy becomes a constant M; thus, the outage performance becomes the same regarding different a and b values, as shown in the high SNR approximation part. Note that the non-linear response will only make sure the harvested energy does not exceed M. In some rare occasions, we can have $P_R^C < P_R^{EH}$, which clearly violates the physical meaning in our model. So these events are excluded from the results.

The outage performance for UE 2 as the function of β is shown in Figure 4.9. The parameters used for this study are $a = 2, \rho_{BR} = 40$ dB. The simulation and analytical results for UE 2 are both presented here and they match well with each other. With the increase of β, the outage probability also increases. The increase slope slows down as β further increases, due to the fact that β is the portion of power assigned to energy harvest unit. The less power remained for transmitting, the higher outage probability it will have. The inconsistence between simulation and analytical results when $\beta = 0.1$ comes from the excluded events when $P_R^C < P_R^{EH}$.

4.2.5 Summary

In the first part of the chapter, we analyze the outage performance of two NOMA relaying schemes. NOMA cooperative scheme needs two time slots to complete one round communication. It uses NOMA in the first time slot and uses DPC precoding in the second time slot for cooperation. NOMA TDMA scheme needs three time slots to complete one round communication. It uses NOMA in the first time slots and then TDMA in the second and third time slots. SIC error propagation is considered in the analysis and the performance degradation is evaluated. The analytical results agree with the simulation results very well. Future work can optimize the power allocation factor α_s and β_s to achieve the best outage performance under different schemes.

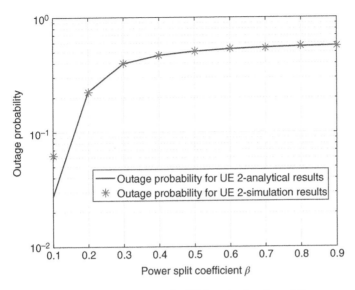

Figure 4.9 Outage performance for UE 2 as the function of β.

In the second part, we consider applying NOMA and D2D relaying in a mmWave-based wireless system that consists of high-power base stations and low-power IoT devices. The lower-power IoT devices do not have external power supplies and have limited battery life. In order to prolong battery life and also to motivate low-power IoT devices to help relay signals from others, low-power IoT devices can harvest energy from electromagnetic signals. To make the energy harvest model more realistic, non-linear energy harvesting model is used. The theoretical analysis on outage probability is given for the proposed scheme and system model. Simulation results validate the accuracy of the analysis.

5

Robust Beamforming in NOMA Cognitive Radio Networks: Bounded CSI

5.1 Background

As a promising technique of improving the spectral efficiency (SE), cognitive radio (CR) techniques have also been investigated for decades, where the secondary users (SUs) may access the spectrum bands of the primary users (PUs), as long as the interference caused by SUs is tolerable [234]. According to [64], in order to implement CR in practice, three operational models have been proposed, namely, opportunistic spectrum access, spectrum sharing, and sensing-based enhanced spectrum sharing. In 5G era, it is envisioned that the combination of NOMA with CR is capable of further improving the SE. As a benefit of its low implementation complexity, spectrum sharing has been widely applied. In [111–119], the authors analyzed the performance of a spectrum sharing CR combined with NOMA. It was shown that the SE can be significantly improved by using NOMA in CR compared to that achieved by using OMA in CR.

On the other hand, the increasing greenhouse gas emissions have become a major concern also in the design of wireless communication networks. According to [81], cellular networks world-wide consume approximately 60 billion kWh energy per year. Moreover, this energy consumption is explosively increasing due to the unprecedented expansion of wireless networks to support ubiquitous coverage and connectivity. Furthermore, because of the rapid proliferation of IoT applications, most battery-driven power-limited IoT devices become useless if their battery power is depleted. Thus it is critical to use energy in an efficient way or to harness renewable energy sources. As remedy, energy harvesting (EH) exploits the pervasive frequency radio signals for replenishing the batteries [118]. There have been two research thrusts on EH using RF technology. One focuses on wirelessly powered networks, where a so-called harvest-then-transmit protocol is applied [22]. The other one uses SWIPT [138–143], which is the focus of this chapter. The contributions of SWIPT in CR have been extensively studied. Specifically, authors of [186] considered the optimal beamforming design in a multiple-input single-output (MISO) CR downlink network. A similar power splitting structure to that of our work is applied at the user side. Hu *et al.* [79], on the other hand, investigated the objective function of EH energy maximization, and a resource allocation problem was formulated to address that goal. Additionally, Mohjazi *et al.* [135] considered the underlay scheme in CR network and proposed the optimal beamforming design. To address both the SE and EE, a MISO NOMA CR using SWIPT is considered

5G and Beyond Wireless Communication Networks, First Edition. Haijian Sun, Rose Qingyang Hu, and Yi Qian.
© 2024 John Wiley & Sons Ltd. Published 2024 by John Wiley & Sons Ltd.

based on a practical non-linear EH model. Robust beamforming design problems are studied under a pair of CSI error models. The related contributions and the motivation of our work are summarized as follows.

5.1.1 Related Work and Motivation

The prior contributions related to this chapter can be divided into two categories based on the EH model adopted, i.e. the linear [138–173, 178, 180–194, 196–225] and the non-linear EH model [22, 24–121, 123–173, 178, 180–194, 196–198]. In the linear EH model, the power harvested increases linearly with the input power, while the EH under the non-linear model exhibits more realistic non-linear characteristics especially at the power-tail.

5.1.1.1 Linear EH Model

In [110], Liu *et al.* analyzed the performance of a cooperative NOMA system relying on SWIPT, which outperformed OMA. Do *et al.* [49] extended [110] and studied the beneficial effect of the user selection scheme on the performance of a cooperative NOMA system using SWIPT. In [219], Yang *et al.* presented a theoretical analysis of two power allocation schemes conceived for a cooperative NOMA system with SWIPT. It was shown that the outage probability achieved under NOMA is lower than that obtained under OMA. Diamantoulakis *et al.* [41] studied the optimal resource allocation design of wireless-powered NOMA systems. The optimal power and time allocation were designed for maximizing the max-min fairness among users. In their following work [42], a joint downlink and uplink scheme was considered in a wireless powered network, followed by comparisons between NOMA and TDMA. The results show that NOMA is more energy efficient in the downlink of SWIPT networks. In order to improve the EE, multiple antennas were applied in a NOMA system associated with SWIPT, and the transmit beamforming and the power splitting factor were jointly optimized for maximizing the transmit rate of users [216].

The contributions in [110–121, 123–173, 178, 180–194, 196–216] investigated conventional wireless NOMA systems, which did not consider the interference between the secondary network and the primary network. Recently, authors of [135, 138] and [238] studied optimal resource allocation problems in CR associated with SWIPT. In [138], an optimal transmit beamforming scheme was proposed in a multi-objective optimization framework. It was shown that there are several tradeoffs in CR-aided SWIPT. Based on the work in [138], the authors proposed a jointly optimal beamforming and power splitting scheme to minimize the transmit power of the base station in multiple-user CR-aided SWIPT [135]. Considering the practical imperfect CSI, Zhou *et al.* [236] studied robust beamforming design problems in MISO CR-aided SWIPT, where the bounded and the gaussian CSI error models were applied. It was shown that the performance achieved under the gaussian CSI error model is better than that obtained under the bounded CSI error model. The work in [236] was then extended to MIMO CR-aided SWIPT in [54] and [213], where the bounded CSI error model was applied in [54] and the gaussian CSI error model was used in [213] and [98]. In contrast to [135–173, 178, 180–194, 196–213], Zhou *et al.* [238] studied robust resource allocation problems in CR-aided SWIPT under opportunistic spectrum access.

5.1.1.2 Non-linear EH Model

In [22], robust resource allocation schemes were proposed for maximizing the sum trans-mission rate or the max-min transmission rate of MIMO-assisted wireless powered com-munication networks, where a practical non-linear EH model is considered. It was shown that a performance gain can be obtained under a practical non-linear EH model over that attained under the linear EH model, see Figure 4.7. In order to maximize the power-efficient and sum-energy harvested by SWIPT systems, Boshkovska *et al.* designed optimal beam-forming schemes in [24] and [20]. Recently, under the idealized perfect CSI assumption, the rate-energy region was quantified in MIMO systems relying on SWIPT and the practi-cal non-linear EH model in [212]. In order to improve the security of a SWIPT system, a robust beamforming design problem was studied under a bounded CSI error model in [23]. The investigations in [22–24] were performed in the context of conventional SWIPT sys-tems. Recently, Wang *et al.* [198] extended a range of classic resource allocation problems into a wireless powered CR counterpart. The optimal channel and power allocation scheme were proposed for maximizing the sum transmission rate.

The resource allocation schemes proposed in [110–121, 123–173, 178, 180–194, 196–216] investigated a conventional NOMA system with SWIPT. The mutual interference should be considered and the QoS of the PUs should be protected in NOMA CR. Moreover, the resource allocation schemes proposed in [135–173, 178, 180–194, 196–238] are based on the classic OMA scheme. Thus, these schemes are not applicable to NOMA CR with SWIPT due to the difference between OMA and NOMA. Furthermore, an idealized linear EH model was applied in [138–173, 178, 180–194, 196–238], which is impractical since the practical power conversion circuit results in a non-linear end-to-end wireless power transfer. Therefore, it is of great importance to design optimal resource allocation schemes for NOMA CR-aided SWIPT based on the practical non-linear EH model.

Although the practical non-linear EH model was applied in [22, 24–121, 123–173, 178, 180–194, 196–198], the authors of [22–24] considered conventional OMA systems using SWIPT. Moreover, the resource allocation scheme proposed in [238] is based on OMA and cannot be directly introduced in NOMA CR-aided SWIPT. However, at the time of writing, there is a scarcity of investigations on robust resource allocation design for NOMA CR-aided SWIPT under the practical non-linear EH model. Several challenges have to be addressed to design robust resource allocation schemes for NOMA CR-aided SWIPT. For example, the impact of the CSI error and of the residual interference due to the imperfect SIC should be considered, which makes the robust resource allocation problem quite challenging. Thus, we study robust resource allocation problems in NOMA CR-aided SWIPT.

5.1.2 Contributions

This chapter expands [22] in three major contexts. Firstly, a NOMA MISO CR-aided SWIPT is considered, while a OMA MIMO wireless powered network was used in [22]. Secondly, the work in [22] relies on the bounded CSI error model, while both the bounded and the gaussian CSI error model are applied in our work. Thirdly, we consider the minimum trans-mit power as the optimization objective, which is not considered in [22].

5.2 System and Energy Harvesting Models

5.2.1 System Model

We consider a downlink CR system with one cognitive base station (CBS), one primary base station (PBS), N PUs, and K SUs [174]. The CBS is equipped with M antennas, while each user and PBS have a single antenna. It is assumed that the SUs are energy-constrained and energy harvest circuits are used. Specifically, the receiver architecture relies on a power splitting design. Once the signal is detected by the receiver, it will be divided into two parts. One part is used for information detection, while the other part for energy harvesting. Similar structures can be found in [110, 216]. To better utilize the radio resources, all UEs are allowed to access the same resource simultaneously. To be specific, the PBS sends messages to all PUs, while the CBS communicates with all SUs simultaneously by applying NOMA principles by controlling the interference from the CBS to PUs below a certain level [111] (Figure 5.1). Let us denote the set of SUs and PUs as $\mathcal{K} = \{1, 2, \dots, K\}$ and $\mathcal{N} = \{1, 2, \dots, N\}$, respectively. The signal received by the kth SU can be expressed as

$$y_k^S = \mathbf{h}_k^\dagger \mathbf{x} + n_k^S, \ k \in \mathcal{K}, \tag{5.1}$$

where $\mathbf{h}_k \in \mathbb{C}^{M \times 1}$ is the channel gain between the CBS and the kth SU, while n_k^S is the joint effect of additive white Gaussian noise (AWGN) and interference from the PBS. $n_k^S \sim \mathcal{CN}(0, \sigma_{k,S}^2)$, where $\sigma_{k,S}^2$ is the power. This interference model represents a worst-case scenario [138]. Furthermore, $\mathbf{x}$ is the message transmitted to SUs after precoding. According to the NOMA principle, we have:

$$\mathbf{x} = \sum_{k=1}^{K} \mathbf{w}_k s_k + \mathbf{v}, \tag{5.2}$$

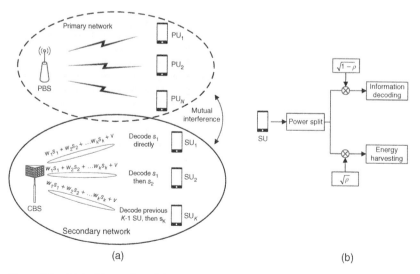

Figure 5.1 (a) An illustration of the system model. (b) The power splitting architecture of SUs.

where $\mathbf{w}_k \in \mathbb{C}^{M\times 1}$ is the precoding vector for the k-th UE and s_k is the corresponding intended message. Furthermore, $\mathbf{v} \in \mathbb{C}^{M\times 1}$ is the energy vector allowing us to improve the energy harvesting efficiency at the SUs. We assume that s_k is unitary, i.e. $\mathbb{E}[|s_k|^2] = 1$, and $\mathbf{v}$ obeys the complex Gaussian distribution, i.e. $\mathbf{v} \sim \mathcal{CN}(\mathbf{0}, \mathbf{V})$, where $\mathbf{V}$ is the covariance matrix of $\mathbf{v}$.

Likewise, the extra interference arriving from the CBS to the n-th PU is

$$y_n^P = \mathbf{g}_n^\dagger \mathbf{x}, \quad n \in \mathcal{N}, \tag{5.3}$$

where $\mathbf{g}_n^\dagger \in \mathbb{C}^{M\times 1}$ is the channel gain between the CBS and the n-th PU [236].

5.2.2 Non-linear EH Model

Most of the existing literature considered an idealized linear energy harvesting model, where the energy collected by the k-th SU is expressed as $E_k^{\text{Linear}} = \eta E_k^{\text{In}}$, $E_k^{\text{In}} = \rho \left(\mathbf{h}_k^\dagger (\sum_{j=1}^K \mathbf{w}_j \mathbf{w}_j^\dagger + \mathbf{V}) \mathbf{h}_k + \sigma_{k,S}^2 \right)$ is the input power, where ρ is the power splitting factor that controls the amount of received energy allocated to energy harvesting, $0 < \rho < 1$, while η is the energy conversion efficiency factor, $0 < \eta \leq 1$. However, measurements relying on real-world testbeds show that a typical energy harvesting model exhibits a non-linear end-to-end characteristic. To be specific, the harvested energy first grows almost linearly with the increase of the input power, and then saturates when the input power reaches a certain level. Several models have been proposed in the literature and one of the most popular ones is [22], which is formulated as follows:

$$E_k^{\text{Practical}} = \frac{\Psi_k^{\text{Practical}} - M_k \Omega_k}{1 - \Omega_k}, \Omega_k = \frac{1}{1 + \exp(a_k b_k)}, \tag{5.4a}$$

$$\Psi_k^{\text{Practical}} = \frac{M_k}{1 + \exp\left(-a_k(E_k^{\text{In}} - b_k)\right)}, \tag{5.4b}$$

where $E_k^{\text{Practical}}$ is the actual energy harvested from the circuit. Furthermore, $\Psi_k^{\text{Practical}}$ represents a function of the input power E_k^{In}. Additionally, M_k is the maximum power that a receiver can harvest, while a_k together with b_k characterizes the physical hardware in terms of its circuit sensitivity, limitations, and leakage currents [22].

On the other hand, the signal received in the k-th SU information decoding circuit is

$$y_k^D = \sqrt{1 - \rho}(\mathbf{h}_k^\dagger \mathbf{x} + n_k^S) + n_k^D, \tag{5.5}$$

where n_k^D is the AWGN imposed by the information decoding receiver.

5.2.3 Bounded CSI Error Model

In this model, we consider a bounded error imposed on the estimated CSI, which can be treated as the worst-case scenario. Specifically, the channels can be modeled as follows.

$$\mathbf{h}_k = \hat{\mathbf{h}}_k + \Delta \mathbf{h}_k, \forall k \in \mathcal{K}, \tag{5.6a}$$

$$\Gamma_k \triangleq \left\{ \Delta \mathbf{h}_k \in \mathbb{C}^{M\times 1} : \|\Delta \mathbf{h}_k\|^2 \leq \varphi_k^2 \right\}, \tag{5.6b}$$

$$\mathbf{g}_n = \hat{\mathbf{g}}_n + \Delta \mathbf{g}_n, \forall n \in \mathcal{N}, \tag{5.6c}$$

$$\Theta_n \triangleq \left\{ \Delta\mathbf{g}_n \in \mathbb{C}^{M\times1} : \|\Delta\mathbf{g}_n\|^2 \leq \psi_n^2 \right\}, \tag{5.6d}$$

where $\hat{\mathbf{h}}_k$ and $\hat{\mathbf{g}}_n$ are the estimated channel vectors for $\mathbf{h}_k$ and $\mathbf{g}_n$, respectively, while Γ_k and Θ_n define the set of channel variations due to estimation errors. The model defines all the uncertainty regions that are confined by power constraints. Furthermore, we use block Rayleigh fading channels, which remain constant within each block, but change from block to block independently.

5.2.3.1 NOMA Transmission

Without loss of generality, we sort the estimated channel of SUs in the ascending order, i.e. $\|\hat{\mathbf{h}}_1\|^2 \leq \|\hat{\mathbf{h}}_2\|^2 \leq \cdots \|\hat{\mathbf{h}}_K\|^2$. According to the SIC principle, SU i can detect and remove SU k's signal, for $1 \leq k < i \leq K$. Thus, when SU i decodes signal s_k, the signals of the previous $(k-1)$ SUs have already been removed from the composite received signal. Due to channel estimation errors, however, these $(k-1)$ signals may not be completely removed, leaving some residual signals as interference. Therefore, the signal at UE i when decoding s_k becomes

$$y_{i,k}^S = \sqrt{1-\rho} \left(\mathbf{h}_i^\dagger \mathbf{w}_k s_k + \sum_{j=1}^{k-1} \Delta\mathbf{h}_i^\dagger \mathbf{w}_j s_j + \sum_{j=k+1}^{K} \mathbf{h}_i^\dagger \mathbf{w}_j s_j + \mathbf{h}_i^\dagger \mathbf{v} + n_k^S \right) + n_k^D$$

Here, the first term is the desired received signal, the second term is the interference due to imperfect channel estimation, and the third term represents the NOMA interference. For notational simplicity, let us denote $\mathbf{W}_k = \mathbf{w}_k \mathbf{w}_k^\dagger$, $\mathbf{V} = \mathbf{v}\mathbf{v}^\dagger$, $S_i^k = \mathbf{h}_i^\dagger \mathbf{W}_k \mathbf{h}_i$, and $T_i^j = \Delta\mathbf{h}_i^\dagger \mathbf{W}_j \Delta\mathbf{h}_i$. The corresponding SINR for the i-th SU after the SIC applied at the receiver is given by:

$$\text{SINR}_i^k = \frac{S_i^k}{\sum_{j=1}^{k-1} T_i^j + \sum_{j=k+1}^{K} S_i^j + \mathbf{h}_i^\dagger \mathbf{V} \mathbf{h}_i + \sigma_{k,S}^2 + \frac{\sigma_D^2}{(1-\rho)}}.$$

Since the signal s_k can be detected at every SU i, as long as $k < i$, there will be a set of SINRs for signal s_k. For CBS, the maximum data rate for SU k should be $R_k = \log_2(1 + \min_{k \leq i \leq K} \text{SINR}_i^k)$. Moreover, the channel estimation error should be considered. The worst-case data rate for SU k becomes

$$R_k = \log_2 \left(1 + \min_{\Delta\mathbf{h}_i \in \Gamma_i} \left\{ \min_{k \leq i \leq K} \text{SINR}_i^k \right\} \right). \tag{5.7}$$

5.3 Power Minimization-Based Problem Formulation

Since $\mathbf{x}$ is a composite signal consisting of all SUs' messages, SIC is applied at the receiver side to detect the received signal. The detection is carried out in the same order of the channel gains, i.e. the SUs with lower channel gain will be decoded first. A pair of imperfect CSI error models are considered, namely a bounded and a gaussian model. We adopt both of these in this chapter and assume that all SUs have a perfect knowledge of their own CSI.

5.3.1 Problem Formulation

In this section, we seek to find the precoding vectors $\mathbf{w}_k$, $k \in \mathcal{K}$, the energy vector $\mathbf{v}$, and the power split ratio ρ, which altogether achieve a satisfactory QoS for all users, and at the same time, they can harvest part of the energy for their future usage. Thus, the problem can be formulated as follows:

$$\mathbf{P}_1 : \min_{\mathbf{W}_k \in \mathbb{C}^{M \times M}, \mathbf{V} \in \mathbb{C}^{M \times M}, \rho} \mathrm{Tr}\left(\sum_{k=1}^{K} \mathbf{W}_k + \mathbf{V} \right) \tag{5.8a}$$

$$\text{s.t.} \ C1 : R_k \geq R_{k,\min} \tag{5.8b}$$

$$C2 : E_k^{\mathrm{Practical}} \geq P_{k,s}, \ \forall \Delta \mathbf{h}_k \in \mathbf{\Gamma}_k, \ \forall k \in \mathcal{K}, \tag{5.8c}$$

$$C3 : \mathbf{g}_n^{\dagger}\left(\sum_{j=1}^{K} \mathbf{W}_j + \mathbf{V} \right) \mathbf{g}_n \leq P_{n,p}, \ \forall \Delta \mathbf{g}_n \in \mathbf{\Theta}_n, \tag{5.8d}$$

$$C4 : \mathrm{Tr}\left(\sum_{k=1}^{K} \mathbf{W}_k + \mathbf{V} \right) \leq P_B, \tag{5.8e}$$

$$C5 : 0 < \rho < 1, \tag{5.8f}$$

$$C6 : \mathbf{V} \succ \mathbf{0}, \mathbf{W}_k \succ \mathbf{0}, \tag{5.8g}$$

$$C7 : \mathrm{Rank}(\mathbf{W}_k) = 1, \ \forall k \in \mathcal{K}. \tag{5.8h}$$

Our goal is to minimize the total transmitted power. The constraint $C1$ ensures that SU k does attain the predefined minimum data rate; $C2$ allows each SU to harvest the amount of energy that at least compensates the static power dissipation $P_{k,s}$; $C3$ is the interference limit for the n-th PU; $C4$ represents the maximum transmit power constraint of the BS; in $C5$, the power split factor should be in the range of $(0, 1)$. The optimization problem $\mathbf{P}_1$ is hard to solve due to its non-convexity constraints $C1$ and $C2$. Moreover, the realistic imperfect CSI imposes another challenge on the original problem. In the following, we transform the variables.

Let us introduce $\gamma_{k,\min} \triangleq (2^{R_{k,\min}} - 1)$. Then $C1$ in (5.8b) becomes

$$\min_{\Delta \mathbf{h}_i \in \mathbf{\Gamma}_i} \frac{S_i^k}{\sum_{j=1}^{k-1} T_i^j + \sum_{j=k+1}^{K} S_i^j + \mathbf{h}_i^{\dagger} \mathbf{V} \mathbf{h}_i + \sigma_{k,S}^2 + \frac{\sigma_D^2}{(1-\rho)}} \geq \gamma_{k,\min}, \tag{5.9}$$

where $i = \{k, k+1, \ldots, K\}$, $\forall k \in \mathcal{K}$. For the notational simplicity, we denote the above constraint as $\Xi_{i,k}$. Thus, $\mathbf{P}_1$ becomes

$$\mathbf{P}_2 : \min_{\mathbf{W}_k \in \mathbb{C}^{M \times M}, \mathbf{V} \in \mathbb{C}^{M \times M}, \rho} \mathrm{Tr}\left(\sum_{k=1}^{K} \mathbf{W}_k + \mathbf{V} \right) \tag{5.10a}$$

$$\text{s.t.} \ C1 : \Xi_{i,k} \tag{5.10b}$$

$$C2 : E_k^{\mathrm{Practical}} \geq P_{k,s}, \ \forall \Delta \mathbf{h}_k \in \mathbf{\Gamma}_k, \ \forall k \in \mathcal{K}, \tag{5.10c}$$

$$C3 : \mathbf{g}_n^{\dagger} \left(\sum_{j=1}^{K} \mathbf{W}_j + \mathbf{V} \right) \mathbf{g}_n \leq P_{n,p}, \forall \Delta \mathbf{g}_n \in \boldsymbol{\Theta}_n, \tag{5.10d}$$

$$(5.8e) - (5.8h). \tag{5.10e}$$

Here, $C6$ comes from the fact that both $\mathbf{V}$ and $\mathbf{W}_k$ are positive semi-definite matrices. The extra constraint that the rank of $\mathbf{W}_k$ should be 1 is also non-convex. In what follows, we first reformulate $C1$ in (5.10b) according to the S-Procedure of [25].

Lemma 5.1 $C1$ *in (5.10b) can be reformulated as*

$$\begin{bmatrix} \alpha_{i,k}\mathbf{I} + \mathbf{C}_k - \gamma_{k,min} \sum_{j=1}^{k-1} \mathbf{W}_j & \mathbf{C}_k \hat{\mathbf{h}}_i \\ \hat{\mathbf{h}}_i^{\dagger} \mathbf{C}_k & -\alpha_{i,k}\varphi_k^2 + \Phi_k \end{bmatrix} \succ \mathbf{0}, \tag{5.11}$$

$\forall k \in \mathcal{K}, i = \{k, k+1, \dots, K\}$, *where* $\mathbf{C}_k = \mathbf{W}_k - \gamma_{k,min}(\sum_{j=k+1}^{K} \mathbf{W}_j + \mathbf{V})$ *and* $\Phi_k = \hat{\mathbf{h}}_i^{\dagger} \mathbf{C}_k \hat{\mathbf{h}}_i - \gamma_{k,min} \left(\sigma_{k,S}^2 + \frac{\sigma_D^2}{(1-\rho)} \right)$, *and* $\alpha_{i,k}$ *is a slack variable conditioned on* $\alpha_{i,k} \geq 0$.

Proof: Given $\mathbf{h}_i = \hat{\mathbf{h}}_i + \Delta \mathbf{h}_i$ and (5.9), we have

$$\Delta \mathbf{h}_i^{\dagger} \left(\gamma_{k,min} \left(\sum_{j \neq k} \mathbf{W}_j + \mathbf{V} \right) - \mathbf{W}_k \right) \Delta \mathbf{h}_i + 2 \, \mathrm{Re} \left\{ \hat{\mathbf{h}}_i^{\dagger} \left(\gamma_{k,min} \left(\sum_{j=k+1}^{K} \mathbf{W}_j + \mathbf{V} \right) - \mathbf{W}_k \right) \Delta \mathbf{h}_i \right\}$$

$$+ 2\hat{\mathbf{h}}_i^{\dagger} \left(\gamma_{k,min} \left(\sum_{j=k+1}^{K} \mathbf{W}_j + \mathbf{V} \right) - \mathbf{W}_k \right) \hat{\mathbf{h}}_i + 2\gamma_{k,min} \left(\sigma_{k,S}^2 + \frac{\sigma_D^2}{(1-\rho)} \right) \leq 0. \tag{5.12}$$

From the fact that $\Delta \mathbf{h}_i^{\dagger} \Delta \mathbf{h}_i - \varphi_k^2 \leq 0$ and according to the S-Procedure, the lemma is proved.

Similarly, $C3$ in (5.10d) can be transformed into

$$\begin{bmatrix} \beta_n \mathbf{I} - \boldsymbol{\Sigma} & -\boldsymbol{\Sigma} \hat{\mathbf{g}}_n \\ -\hat{\mathbf{g}}_n^{\dagger} \boldsymbol{\Sigma} & -\beta_n \psi_n^2 - \hat{\mathbf{g}}_n^{\dagger} \boldsymbol{\Sigma} \hat{\mathbf{g}}_n + P_{n,p} \end{bmatrix} \succ \mathbf{0}, \forall n \in \mathcal{N}, \tag{5.13}$$

where $\boldsymbol{\Sigma} = \sum_{j=1}^{K} \mathbf{W}_j + \mathbf{V}$, and $\beta_n \geq 0$ is also a slack variable.

Next, we apply similar manipulations to 5.10c, which becomes

$$\min_{\Delta \mathbf{h}_k \in \Gamma_k} \rho \left(\mathbf{h}_k^{\dagger} \boldsymbol{\Sigma} \mathbf{h}_k + \sigma_{k,S}^2 \right) \geq D_k, \tag{5.14}$$

where $D_k = -\ln \left(\frac{1}{P_{k,s}(1-\Omega_k)/M_k + \Omega_k} - 1 \right) / a_k + b_k$ is a constant. This condition holds, provided that $a_k > 0$, which is always true in real systems.

Then, applying the S-Procedure to (5.14), we have the following

$$\begin{bmatrix} \theta_k \mathbf{I} + \boldsymbol{\Sigma} & \boldsymbol{\Sigma} \hat{\mathbf{h}}_k \\ \hat{\mathbf{h}}_k^{\dagger} \boldsymbol{\Sigma} & -\theta_k \varphi_k^2 + \hat{\mathbf{h}}_k^{\dagger} \boldsymbol{\Sigma} \hat{\mathbf{h}}_k + \sigma_{k,S}^2 - \frac{D_k}{\rho} \end{bmatrix} \succ \mathbf{0}, \tag{5.15}$$

$\forall k \in \mathcal{K}$, where $\theta_k \geq 0$.

Therefore, $\mathbf{P}_2$ becomes

$$\mathbf{P}_3 : \min_{\mathbf{W}_k, \mathbf{V}, \rho, \{\alpha_{i,k}\}, \{\beta_n\}, \{\theta_k\}} \mathrm{Tr} \left(\sum_{k=1}^{K} \mathbf{W}_k + \mathbf{V} \right) \tag{5.16a}$$

s.t.   (5.11), (5.13), (5.15), (5.8e), (5.8f), (5.8g),   (5.16b)

$$\alpha_{i,k}, \beta_n, \theta_k \geq 0,$$

$$\forall k \in \mathcal{K}, i = \{k, k+1, \ldots, K\}, \forall n \in \mathcal{K}. \tag{5.16c}$$

Observe that we drop (5.8h), since it is not a convex term. This relaxation is commonly referred to as the semi-definite relaxation (SDR) technique. For the specific problem in $\mathbf{P}_2$, the following theorem proves that the optimal $\mathbf{W}_k$ has a limited rank.

Theorem 5.1   *If $\mathbf{P}_2$ is feasible, the rank of $\mathbf{W}_k, k \in \mathcal{K}$ is always less than or equal to 2.*

Proof: See Appendix A.

The transformed problem $\mathbf{P}_3$ is not convex because of the coupling variables ρ in (5.15) and $(1 - \rho)$ in the denominator of (5.11). To be able to take advantage of the *CVX* software package, we introduce a pair of auxiliary variables. Specifically, let $p = \frac{1}{1-\rho}$ and $q = \frac{1}{\rho}$. In this way, (5.11), (5.13), and (5.15) become convex terms. Then, we have additional constraints for p and q:

$$p \geq \frac{1}{1 - \rho} \text{ and } q \geq \frac{1}{\rho}. \tag{5.17}$$

It may be readily verified that this transformation does not change the optimal solution of $\mathbf{P}_3$.

5.3.2 Matrix Decomposition

Now we proceed to find the solution of the problem $\mathbf{P}_2$, after which there is one more step to get the original solution for $\mathbf{w}_k$. If $\mathbf{W}_k$ yields rank 1, we can simply write $\mathbf{W}_k^\star = \mathbf{w}_k^\star \mathbf{w}_k^{\star\dagger}$. Otherwise, if Rank($\mathbf{W}_k^\star$) = 2, we have several optional approaches to extract $\mathbf{w}_k^\star$. To name a few, we list two methodologies here.

1. *Eigen-decomposition.* Let us denote two eigenvalues of $\mathbf{W}_k^\star$ by λ_1 and λ_2, where $\lambda_1 > \lambda_2 \geq 0$. Clearly, $\mathbf{W}_k^\star = \lambda_1 \mathbf{w}_{1k} \mathbf{w}_{1k}^\dagger + \lambda_2 \mathbf{w}_{2k} \mathbf{w}_{2k}^\dagger$, $\mathbf{w}_{ik}, i = \{1, 2\}$ are the corresponding eigenvectors. To get the rank 1 approximation from a rank 2 matrix, we can let the solution of the original problem be $\hat{\mathbf{w}}_k = \sqrt{\lambda_1} \mathbf{w}_{1k} \mathbf{w}_{1k}^\dagger$, provided it is feasible.
2. *Randomization technique.* Similar to eigen-decomposition, we first decompose $\mathbf{W}_k^\star$ according to $\mathbf{W}_k^\star = \mathbf{U}_k \mathbf{T}_k \mathbf{U}_k^\dagger$. Then, we let $\hat{\mathbf{w}}_k = \mathbf{U}_k \mathbf{T}_k^{1/2} \mathbf{e}_k$, where the m-th element of $\mathbf{e}_k$ is $[\mathbf{e}_k]_m = e^{j\theta_{k,m}}$ and $\theta_{k,m}$ obeys an independent and uniform distribution within $[0, 2\pi)$.

The above two methods are essentially the same. If we want to get a more precise result, another scaling factor can be added. Specifically, let us define c_k as the scaling factor yet to be determined. Certainly, the problem can be transformed in terms of $\mathbf{W}_k$ and c_k, once we get the optimal value, we can apply either one of the above methods to get a better result. Another point worth noting here is that when the rank of $\mathbf{W}_k$ is 2, there only exists the approximation result of $\mathbf{w}_k^\star$, and this approximation always provides an upper bound.

5.4 Maximum Harvested Energy Problem Formulation

In contrast to Section 5.3, where the minimum transmission power problem is considered, in the following we consider the optimization problem of maximizing the total harvested energy. This problem has important real-world applications, since most of the consumer electronics products are battery-driven and thus their energy efficiency is critical. In this section, we first formulate the problem, then we transform it in a convex way so that an existing software package can solve it efficiently. A one-dimensional search algorithm will be used.

Upon considering the imperfect CSI model used in (5.6), the maximum total harvested energy of all SUs can be formulated as follows:

$$
\mathbf{P_6}: \max_{\substack{\mathbf{W}_k \in \mathbb{C}^{M \times M}, \mathbf{V} \in \mathbb{C}^{M \times M} \\ \rho, \{\alpha_{i,k}\}, \{\beta_n\}, \{\theta_k\}}} \sum_{k=1}^{K} E_k^{\text{Practical}} \tag{5.18a}
$$

s.t. (5.11), (5.13), (5.8e), (5.8f), (5.8g), (5.8h), $\qquad$ (5.18b)

$$
\alpha_{i,k}, \beta_n, \theta_k \geq 0, \forall k \in \mathcal{K}, i = \{k, k+1, \dots, K\}, \forall n \in \mathcal{K}. \tag{5.18c}
$$

The rank operation is not convex; thus, we drop the constraint (5.8h) first, as previously in $\mathbf{P_3}$. Additionally, the objective function relies on a realistic non-linear energy harvesting model, and it is not convex either. Essentially, it is a sum-of-ratio problem, and its global optimization is possible by applying the following transformations:

$$
\max_{\substack{\mathbf{W}_k \in \mathbb{C}^{M \times M}, \mathbf{V} \in \mathbb{C}^{M \times M} \\ \rho, \{\alpha_{i,k}\}, \{\beta_n\}, \{\theta_k\}, \{\tau_k\}}} \sum_{k=1}^{K} \frac{M_k}{1 + \exp\left(-a_k(\tau_k - b_k)\right)} \tag{5.19a}
$$

$$
E_k^{\text{In}} \geq \tau_k, \forall \Delta \mathbf{h}_k. \forall k \in \mathcal{K}. \tag{5.19b}
$$

After applying the $\mathcal{S}$-Procedure of [25] to (5.19b), it becomes

$$
\begin{bmatrix} \theta_k \mathbf{I} + \Sigma & \Sigma \hat{\mathbf{h}}_k \\ \hat{\mathbf{h}}_k^\dagger \Sigma & -\theta_k \varphi_k^2 + \hat{\mathbf{h}}_k^\dagger \Sigma \hat{\mathbf{h}}_k + \sigma_{k,S}^2 - \frac{\tau_k}{\rho} \end{bmatrix} \succ \mathbf{0}, \tag{5.20}
$$

$\forall k \in \mathcal{K}$. Furthermore, according to [24, 92], if $\mathbf{P_6}$ has the optimal solutions $\mathbf{W}_k^\star$ and $\mathbf{V}^\star$, there exist two sets of vectors $\mu = \{\mu_1, \mu_2, \dots, \mu_K\}$ and $\epsilon = \{\epsilon_1, \epsilon_2, \dots, \epsilon_K\}$ such that the solutions are also optimal for the following equivalent parametric optimization problem:

$$
\mathbf{P_7}: \max_{\substack{\mathbf{W}_k \in \mathbb{C}^{M \times M} \\ \mathbf{V} \in \mathbb{C}^{M \times M} \\ \rho, \{\alpha_{i,k}\}, \{\beta_n\}, \{\theta_k\}, \{\tau_k\}}} \sum_{k=1}^{K} \mu_k \left\{ M_k - \epsilon_k \left(1 + \exp(-a_k(\tau_k - b_k))\right) \right\}. \tag{5.21}
$$

The optimal solutions and the vectors should satisfy

$$
\epsilon_k \left(1 + \exp(-a_k(\tau_k^\star - b_k))\right) - M_k = 0, \tag{5.22a}
$$

$$
\mu_k \left(1 + \exp(-a_k(\tau_k^\star - b_k))\right) - 1 = 0, \forall k \in \mathcal{K}, \tag{5.22b}
$$

where $E_k^{\text{In},\star} = \rho^\star \left(\mathbf{h}_k^\dagger (\sum_{j=1}^K \mathbf{W}_j^\star + \mathbf{V}^\star) \mathbf{h}_k + \sigma_{k,S}^2 \right) \geq \tau_k^\star$.

Now, the objective function has the *log-concave* form and it can be solved given the sets μ and ϵ. The iterative update of the vector sets can be carried out in the following way. Let us define the function $\mathcal{F}(\mu, \epsilon) = \left[\epsilon_k \left(1 + \exp(-a_k(\tau_k^\star - b_k)) \right) - M_k, \ldots, \mu_k \left(1 + \exp(-a_k(\tau_k^\star - b_k)) \right) - 1 \right], \forall k \in \mathcal{K}$. The next set of values of μ and ϵ can be updated by solving $\mathcal{F}(\mu, \epsilon) = 0$. Specifically, in the q-th iteration, we update them as:

$$\mu^{q+1} = \mu^q + \varpi^q \mathbf{p}^q, \ \epsilon^{q+1} = \epsilon^q + \varpi^q \mathbf{p}^q, \tag{5.23}$$

where $\mathbf{p}^q = [\mathcal{F}'(\mu, \epsilon)]\mathcal{F}(\mu, \epsilon)$, $\mathcal{F}'(\mu, \epsilon)$ is the Jacobian matrix of $\mathcal{F}(\mu, \epsilon)$, ϖ^q is the largest ϖ^l that satisfies $\| \mathcal{F}(\mu^q + \varpi^l \mathbf{p}^q, \epsilon^q + \varpi^l \mathbf{p}^q) \| \leq (1 - t\varpi^l) \| \mathcal{F}(\mu, \epsilon) \|, l = 1, 2, \ldots, 0 < \varpi^l < 1$, and $0 < t < 1$ [24, 92].

A two-loop algorithm is proposed for solving the problem. The outer loop gives μ and ϵ as the inputs of the inner loop, while the inner loop finds $\mathbf{W}_k^\star$ and $\mathbf{V}^\star$. Observe that in (5.20), there is a coupling variable $\frac{\tau_k}{\rho}$, which is convex with a given ρ. Therefore, in the inner loop, we have to perform a one-dimensional search for ρ as well. The detailed algorithm is formulated in **Algorithm 5.1**.

Algorithm 5.1 Robust Precoding Design for EH Maximization Problem

---

1: **Input**: Minimum required data rate R_k of SU k, noise power $\sigma_{k,S}^2$ and σ_D^2, channel uncertainty φ_k^2 and ψ_k^2, maximum allowed interference power $P_{n,p}$ for PU n, maximum BS transmitted power P_B, and randomly generated estimated channel $\hat{\mathbf{h}}_k$ and $\hat{\mathbf{g}}_n$.

2: **Initialization**: Iteration number $q = 0, p = 1$, initial value of ρ as ρ_{start}, step s, end value ρ_{end}, μ^0, and ϵ^0, loop stop criteria m_{th}.

3: **One-dimensional Search**:

4: **for** $\rho = \rho_{start} : s : \rho_{end}$ **do**

5:     **repeat**: {Outer Loop}

6:         Solve for the optimization problem $\mathbf{P}_7$: {Inner Loop}

7:         **if** ($\mathbf{P}_7$ is feasible) **then**

8:             Obtain $\mathbf{W}_k^q$ and $\mathbf{V}^q$.

9:         **else**

10:             Break from the outer loop.

11:         **end if**

12:         Update μ^{q+1} and ϵ^{q+1} according to (5.23), then let $q = q + 1$.

13:     **until** $\left| \mu_k^{q+1} \left\{ M_k - \epsilon_k^{q+1} \left(1 + \exp(-a_k(\tau_k - b_k)) \right) \right\} \right| < m_{th}$

14:     Calculate $E_{sum}^i = \sum_k E_k^{Practical}$, then let $i = i + 1, q = 0$.

15: **end for**

16: Find the maximum value among all E_{sum}^i, and the precoding and energy matrix.

17: **Output**: Use either of the methods to get the precoding vector $\mathbf{w}_k^{opt}$ and $\mathbf{V}^{opt}$.

5.4.1 Complexity Analysis

For the CBS power minimization problem under the bounded CSI model, $\mathbf{P}_3$ has $\frac{K(K+1)}{2}$ linear matrix inequality (LMI) constraints of size $(M + 1)$ in (13) due to the higher decoding complexity. Furthermore, we have N LMI constraints of size $(M + 1)$ in (15) and K LMI constraints of size $(M + 1)$ in (17). Additionally, in (12g), there are $(K + 1)$ LMI constraints associated with size M, and a total of $\frac{K(K+1)}{2} + 2N + K + 2$ linear constraints. Thus, according to [236] and the reference therein, the total complexity becomes

$$C_{com}^{B} = \ln(\tau^{-1})n\sqrt{\Psi_{comp}^{1}}\left(\left(\frac{K(K+1)}{2} + N + 2K + 1\right)[(M + 1)^3 + n(M + 1)^2]\right.$$

$$+ (K + 1)(M^3 + nM^2) + \frac{K(K+1)}{2} + 2N + K + 2 + n^2\Bigg), \tag{5.24}$$

where $n = \mathcal{O}\left((K + 1)M^2 + N + K + \frac{K(K+1)}{2}\right)$, $\mathcal{O}$ is the big-O notation. Furthermore, we have $\Psi_{comp}^{1} = (\frac{K(K+1)}{2} + N + 2K + 1)M + K^2 + 4N + 3K + 4$, and τ is the accuracy of iteration.

For the maximum harvested energy problem, with bounded channel model, since the difference with that of power minimization problem is that a maximum of T_{max} number of iterations will be performed for one-dimensional search. Hence, the complexity is $T_{max} C_{com}^{B}$.

5.5 Numerical Results

In this section, we present our simulation results for characterizing the performance of the proposed robust beamforming conceived with NOMA under the bounded estimation error models. Unless otherwise stated, the parameters are chosen as in Table 5.1.

5.5.1 Power Minimization Problem

Figure 5.2 shows the empirical CDFs of the minimum transmit power of the CBS for both the imperfect CSI estimation error models. The maximum power P_B is set to 2 Watts. For comparison, we also include the case of OMA, since it represents the traditional access technology. Observe that in order to reduce the inter-user interference, each OMA user relies exclusively on a single time slot. Thus, a total of K time slots are required instead of a single one in our scheme. To make a fair comparison, each SU's achievable data rate should be averaged over all K time slots, which becomes $R_k^{OMA} = \frac{1}{K}\log_2(1 + SINR_k^{OMA})$. Reduced

Table 5.1 Simulation parameters.

Parameters	Values
Number of SUs and PUs	$K = 3, N = 2$
Noise powers	$\sigma_{k,S}^2 = 0.1$, $\sigma_D^2 = 0.01$
Minimum required EH power	$P_{k,s} = 0.01$ Watt
Maximum tolerable interference of PUs	$P_{n,p} = -18$ dBm
Estimated channel gains	$\hat{\mathbf{h}}_k \sim \mathcal{CN}(\mathbf{0}, 0.8\mathbf{I})$ $\hat{\mathbf{g}}_n \sim \mathcal{CN}(\mathbf{0}, 0.1\mathbf{I})$
Outage probability threshold	$\xi_k = \xi_{k,s} = \xi_{n,p} = 0.05$
Non-linear EH model	$M_k = 24$ mW, $a_k = 150$ $b_k = 0.014$ [24]

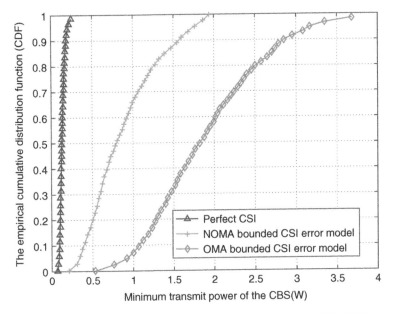

Figure 5.2 The empirical CDF of the minimum transmit power of the CBS under different channel conditions. CBS antenna number $M = 10$, $P_B = 2$ Watts, $R_{min} = 1$ bit/s/Hz.

interference is achieved at the cost of a lower spectral and energy efficiency. We also observe that under both channel error models, the performance of NOMA is better than that of OMA. This is because for OMA, the lower spectral efficiency makes the SU data rate requirement harder to be satisfied. Hence the CBS has to apply a higher transmission power to compensate for that, which leads to a much higher energy consumption. Figure 5.2 is generated from 1,000 independent realizations of different channel conditions. As expected, the performance under perfect CSI is the best, since no additional power is used to compensate for the channel uncertainties.

Figure 5.3 shows the minimum transmit power of the CBS as a function of the minimum required SNR of SUs, $\gamma_{k,min}$. As the SNR increases, the power increases under all CSI cases. Also, perfect CSI requires the least power, followed by NOMA in the bounded CSI model and OMA bounded CSI model. Besides, compared to OMA, the CBS power in NOMA grows slowly. In the parameter setting, $\gamma_{k,min}$ plays a more important role in the constraints. For $\gamma_{k,min} = 2$ in the NOMA case, the equivalent SNR for OMA will be 26. Thus, the gap between OMA and NOMA further increases with the required SNR.

The impact of the CBS antenna number is illustrated in Figure 5.4, where the performance with different CBS antenna numbers and channel uncertainties is plotted. Specifically, Figure 5.4 illustrates how the number of antennas affects the overall performance. The power required increases, when the SNR of SUs grows, regardless of how many antennas are mounted at the CBS. It is also observed that the minimum power required decreases when the number of antennas increases, since a larger number of antennas result in a higher degree of freedom (DoF). Clearly, channel estimation error affects the bounded CSI scenario (Figure 5.5).

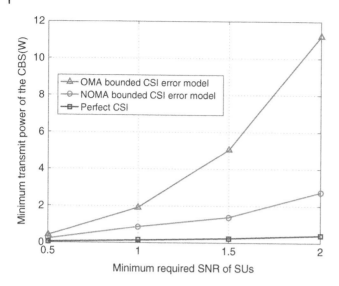

Figure 5.3 The minimum transmit power of the CBS vs. the required SNR of SUs for $M = 10$, $P_B = 8$ Watts.

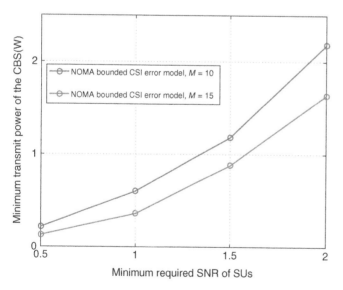

Figure 5.4 Impact of the number of CBS antennas on the minimum transmitted power required in two imperfect CSI scenarios, $M = 15$, $R_{min} = 1$ bit/s/Hz, $P_B = 8$ Watts.

5.5.2 Energy Harvesting Maximization Problem

In this section, we present results for the maximum EH as our objective function. The CBS power is $P_B = 2$ Watts. Figure 5.6 characterizes the average maximum EH power vs. the interference tolerated by the PUs. One can observe that the energy harvested monotonically increases, when the maximum interference tolerated by the PUs grows, where a higher

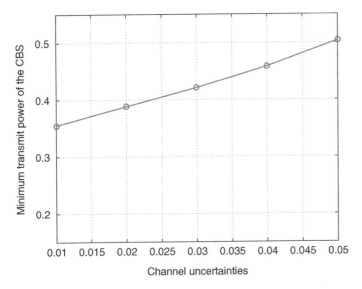

Figure 5.5 Impact of channel uncertainties ψ_n and φ_k on the overall minimum transmit power of the CBS, $M = 15$, $R_{min} = 1$ bit/s/Hz, $P_B = 8$ Watts.

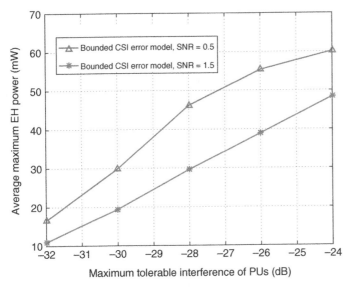

Figure 5.6 Average maximum EH power under different interferences tolerated by the PUs, $M = 10$.

$P_{n,p}$ allows for a larger transmission power, leading to the increase of the harvested energy. When the channel conditions are better, less power is required for satisfying the data rate requirements. Hence more power can be reserved for EH. This also explains that when the required SNR is low, a high EH power can be achieved.

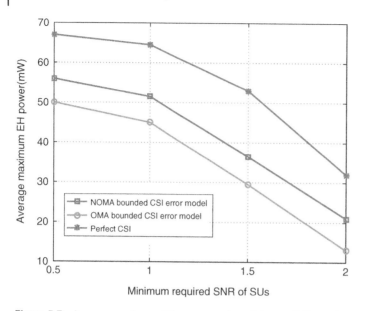

Figure 5.7 Average maximum EH power vs. the minimum SNR required by the SUs, $M = 10$.

The impact of minimum SNRs required by the SUs is illustrated in Figure 5.7. The number of CBS antennas is $M = 10$ and the interference threshold $P_{n,p}$ is set to -24 dBm. We also list the results for the OMA cases. As expected, we can see that the maximum EH power decreases significantly when the SNR grows. This is because more power has to be used for information detection, which leaves less power for energy harvesting.

Figure 5.8 shows the average total EH power vs. the number of SUs. It can be observed that the total EH power grows, when the number of SUs increases, since more nodes participate

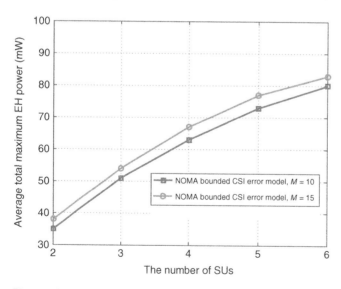

Figure 5.8 Average total EH power vs. the number of SUs for $P_{n,p} = -24$ dBm, $r_{min} = 1$ bit/s/Hz.

in the harvesting process. Additionally, we can see that when the number of antennas is higher, more EH power can be achieved. This is because more antennas give a higher system DoF; therefore, less power is sufficient for information detection.

5.6 Summary

In this chapter, we considered MISO-NOMA CR-aided SWIPT under the bounded CSI estimation error model. A non-linear EH model was applied. Robust beamforming and power splitting control were jointly designed for achieving the minimum transmission power and maximum EH. We transformed the non-convex minimum transmission power optimization problems into a convex form while applying a one-dimensional search algorithm to solve the maximum EH problem. Our simulation results showed that the performance achieved by using NOMA is better than that obtained by using the traditional OMA.

6

Robust Beamforming in NOMA Cognitive Radio Networks: Gaussian CSI

6.1 Gaussian CSI Error Model

In Chapter 5, we introduced a bounded channel model that defines a confined region for the channel variations. It provides a worst-case estimation. Another commonly used more realistic estimation model assumes that the channel estimation error obeys the Gaussian distribution [197, 225, 236], which is formulated as follows:

$$\mathbf{h}_k = \hat{\mathbf{h}}_k + \Delta\mathbf{h}_k, \ \Delta\mathbf{h}_k \sim \mathcal{CN}(\mathbf{0}, \mathbf{H}_k), \ \forall k \in \mathcal{K}, \tag{6.1a}$$

$$\mathbf{g}_n = \hat{\mathbf{g}}_n + \Delta\mathbf{g}_n, \ \Delta\mathbf{g}_n \sim \mathcal{CN}(\mathbf{0}, \mathbf{G}_n), \ \forall n \in \mathcal{N}, \tag{6.1b}$$

where $\Delta\mathbf{h}_k$ and $\Delta\mathbf{g}_n$ are the channel estimation error vectors, while $\hat{\mathbf{h}}_k$ and $\hat{\mathbf{g}}_n$ are the channel vectors estimated at the BS side. Furthermore, $\mathbf{H}_k$ and $\mathbf{G}_n$ are the covariance matrices of the estimation error vectors.

6.2 Power Minimization-Based Problem Formulation

Even though we use a different channel model, the residual interference due to imperfect CSI estimation affects the message detection similarly to the bounded error model. Thus the achievable data rate expression of SU k remains the same except that $\Delta\mathbf{h}_k$ is in a new set. In contrast to the existing NOMA contributions on imperfect CSI [172], in this chapter we use the above-mentioned Gaussian estimation error model to form an optimization problem as follows [174]:

$$\mathbf{P}_1 : \min_{\mathbf{W}_k \in \mathbb{C}^{M \times M}, \mathbf{V} \in \mathbb{C}^{M \times M}, \rho} \ \mathrm{Tr}\left(\sum_{k=1}^{K} \mathbf{W}_k + \mathbf{V}\right) \tag{6.2a}$$

$$\text{s.t. } C1 : \ \Pr\{R_k \geq R_{k,\min}\} \geq 1 - \xi_k, \ \forall k \in \mathcal{K}, \tag{6.2b}$$

$$C2 : \ \Pr\{E_k^{\text{Practical}} \geq P_{k,s}\} \geq 1 - \xi_{k,s}, \tag{6.2c}$$

5G and Beyond Wireless Communication Networks, First Edition. Haijian Sun, Rose Qingyang Hu, and Yi Qian.
© 2024 John Wiley & Sons Ltd. Published 2024 by John Wiley & Sons Ltd.

$$\forall \Delta \mathbf{h}_k \sim \mathcal{CN}(\mathbf{0}, \mathbf{H}_k), \ \forall k \in \mathcal{K},$$

$$C3: \ \Pr\left\{\mathbf{g}_n^\dagger \Sigma \mathbf{g}_n \le P_{n,p}\right\} \ge 1 - \xi_{n,p}, \tag{6.2d}$$

$$\forall \Delta \mathbf{g}_n \sim \mathcal{CN}(\mathbf{0}, \mathbf{G}_n), \ \forall n \in \mathcal{N},$$

$$C4: \ \mathrm{Tr}\left(\sum_{k=1}^{K} \mathbf{W}_k + \mathbf{V}\right) \le P_B, \tag{6.2e}$$

$$C5: \ 0 < \rho < 1, \tag{6.2f}$$

$$C6: \ \mathbf{V} > \mathbf{0}, \mathbf{W}_k > \mathbf{0}, \tag{6.2g}$$

$$C7: \ \mathrm{Rank}(\mathbf{W}_k) = 1, \ \forall k \in \mathcal{K}. \tag{6.2h}$$

Similar to Chapter 5, we seek to find the precoding vectors $\mathbf{w}_k$, $k \in \mathcal{K}$, the energy vector $\mathbf{v}$, and the power split ratio ρ, which altogether achieve a satisfactory QoS for all users, and at the same time, they can harvest part of the energy for their future usage. Here, we assume that the probability of having a rate of R_k is higher than $R_{k,\min}$, which is a predefined value, and we use the threshold ξ_k to control the probability. Likewise, $\xi_{k,s}$ and $\xi_{n,p}$, where $k \in \mathcal{K}$ and $n \in \mathcal{N}$, are used for controlling the outage probability of harvested energy of the kth SU and the interference experienced by the n-th PU, respectively. $\mathbf{P}_1$ is hard to solve owing to its non-convexity, together with constraints $C1 - C3$, which involve probability and uncertainty. Inspired by Zhou *et al.* [236], we solve the resulted optimization problem with the aid of approximations by applying Bernstein-type inequalities [194].

6.2.1 Bernstein-Type Inequality I

Let $f(\mathbf{z}) = \mathbf{z}^\dagger \mathbf{A} \mathbf{z} + 2\mathrm{Re}\{\mathbf{z}^\dagger \mathbf{b}\} + c$, where $\mathbf{A} \in \mathbb{H}^N$, $\mathbf{b} \in \mathbb{C}^{N \times 1}$, $c \in \mathbb{R}$, and $\mathbf{z} \sim \mathcal{CN}(\mathbf{0}, \mathbf{I})$. For any $\xi \in (0, 1]$, an approximate and convex form of [194]

$$\Pr\{f(\mathbf{z}) \ge 0\} \ge 1 - \xi, \tag{6.3}$$

can be written as

$$\mathrm{Tr}(\mathbf{A}) - \sqrt{-2\ln(\xi)}v_1 + \ln(\xi)v_2 + c \ge 0, \tag{6.4a}$$

$$\left\| \begin{bmatrix} \mathrm{vec}(\mathbf{A}) \\ \sqrt{2}\mathbf{b} \end{bmatrix} \right\| \le v_1, \tag{6.4b}$$

$$v_2 \mathbf{I} + \mathbf{A} > \mathbf{0}, v_2 \ge 0. \tag{6.4c}$$

Here, v_1 and v_2 are slack variables.

In order to use the above lemma, we have to transform $\Delta \mathbf{h}_i$ to a standard complex Gaussian vector. Let $\Delta \mathbf{h}_i = \mathbf{H}_i^{1/2} \tilde{\mathbf{h}}_i$, where $\tilde{\mathbf{h}}_i \sim \mathcal{CN}(\mathbf{0}, \mathbf{I})$. Substituting it into (5.9), the convex approximation becomes

$$\mathrm{Tr}\left(\mathbf{H}_i^{1/2}(\mathbf{C}_k - \gamma_{k,\min} \sum_{j=1}^{k-1} \mathbf{W}_j)\mathbf{H}_i^{1/2}\right) - \sqrt{-2\ln(\xi_k)}v_{1i,k} + \ln(\xi_k)v_{2i,k} + c_{i,k} \ge 0, \tag{6.5a}$$

$$c_{i,k} = \hat{\mathbf{h}}_i^\dagger \mathbf{C}_k \hat{\mathbf{h}}_i - r_{k,\min}\left(\sigma_{k,S}^2 + \frac{\sigma_D^2}{1-\rho}\right), \tag{6.5b}$$

$$\left\|\begin{bmatrix} \mathrm{vec}\left(\mathbf{H}_i^{1/2}(\mathbf{C}_k - \gamma_{k,\min}\sum_{j=1}^{k-1}\mathbf{W}_j)\mathbf{H}_i^{1/2}\right) \\ \sqrt{2}\mathbf{H}_i^{1/2}\mathbf{C}_k\hat{\mathbf{h}}_i \end{bmatrix}\right\| \le v_{1i,k}, \tag{6.5c}$$

$$v_{2i,k}\mathbf{I} + \left(\mathbf{H}_i^{1/2}(\mathbf{C}_k - \gamma_{k,\min}\sum_{j=1}^{k-1}\mathbf{W}_j)\mathbf{H}_i^{1/2}\right) \succ \mathbf{0},$$

$$v_{2i,k} \ge 0, \forall k \in \mathcal{K}, i = \{k, \ldots, K\}, \tag{6.5d}$$

where $v_{1i,k}$ and $v_{2i,k}$ are slack variables.

For (6.2d), we use a simple transformation similar as that in (5.14), which leads to:

$$\mathrm{Pr}\left\{\rho(\mathbf{h}_k^\dagger \mathbf{\Sigma}\mathbf{h}_k + \sigma_{k,S}^2) \ge D_k\right\} \ge 1 - \xi_{k,s}. \tag{6.6}$$

Furthermore, by applying the inequalities in (6.4), (6.6) can be expressed as

$$\mathrm{Tr}\left(\mathbf{H}_k^{1/2}\mathbf{\Sigma}\mathbf{H}_k^{1/2}\right) - \sqrt{-2\ln(\xi_{k,s})}v_{1k,s} + \ln(\xi_{k,s})v_{2k,s} + c_{k,s} \ge 0, \tag{6.7a}$$

$$c_{k,s} = \hat{\mathbf{h}}_k^\dagger \mathbf{\Sigma}\hat{\mathbf{h}}_k + \sigma_{k,S}^2 - \frac{D_k}{\rho}, \tag{6.7b}$$

$$\left\|\begin{bmatrix} \mathrm{vec}\left(\mathbf{H}_k^{1/2}\mathbf{\Sigma}\mathbf{H}_k^{1/2}\right) \\ \sqrt{2}\mathbf{H}_k^{1/2}\mathbf{\Sigma}\hat{\mathbf{h}}_k \end{bmatrix}\right\| \le v_{1k,s}, \tag{6.7c}$$

$$v_{2k,s}\mathbf{I} + \left(\mathbf{H}_k^{1/2}\mathbf{\Sigma}\mathbf{H}_k^{1/2}\right) \succ \mathbf{0}, v_{2k,s} \ge 0, \forall k \in \mathcal{K}, \tag{6.7d}$$

where $v_{1k,s}$ and $v_{2k,s}$, $k \in \mathcal{K}$, are slack variables.

6.2.2 Bernstein-Type Inequality II

Let $f(\mathbf{z}) = \mathbf{z}^\dagger \mathbf{A}\mathbf{z} + 2\mathrm{Re}\{\mathbf{z}^\dagger \mathbf{b}\} + c$, where $\mathbf{A} \in \mathbb{H}^N$, $\mathbf{b} \in \mathbb{C}^{N\times 1}$, $c \in \mathbb{R}$, and $\mathbf{z} \sim \mathcal{CN}(\mathbf{0}, \mathbf{I})$. For any $\xi \in (0, 1]$, an approximate and convex form for [13]

$$\mathrm{Pr}\{f(\mathbf{z}) \le 0\} \ge 1 - \xi, \tag{6.8}$$

can be written as

$$\mathrm{Tr}(\mathbf{A}) + \sqrt{-2\ln(\xi)}v_1 - \ln(\xi)v_2 + c \ge 0, \tag{6.9a}$$

$$\left\|\begin{bmatrix} \mathrm{vec}(\mathbf{A}) \\ \sqrt{2}\mathbf{b} \end{bmatrix}\right\| \le v_1, \tag{6.9b}$$

$$v_2\mathbf{I} - \mathbf{A} \succ \mathbf{0}, v_2 \ge 0, \tag{6.9c}$$

where v_1 and v_2 are slack variables.

We apply Bernstein-type Inequality II to (6.2e), and let $\Delta\mathbf{g}_n = \mathbf{G}_n^{1/2}\tilde{\mathbf{g}}_n$, where $\tilde{\mathbf{g}}_n \sim \mathcal{CN}(\mathbf{0}, \mathbf{I})$ is a standard Gaussian vector. We can have the following convex-form approximation.

$$\mathrm{Tr}(\mathbf{G}_n^{1/2}\mathbf{\Sigma}\mathbf{G}_n^{1/2}) + \sqrt{-2\ln(\xi_{n,p})}v_{1,n} - \ln(\xi_{n,p})v_{2,n} + c_n \ge 0, \tag{6.10a}$$

$$c_n = \hat{\mathbf{g}}_n^\dagger \mathbf{\Sigma} \hat{\mathbf{g}}_n - P_{n,p}, \tag{6.10b}$$

$$\left\| \begin{bmatrix} \text{vec}(\mathbf{G}_n^{1/2}\mathbf{\Sigma}\mathbf{G}_n^{1/2}) \\ \sqrt{2}\mathbf{G}_n^{1/2}\mathbf{\Sigma}\hat{\mathbf{g}}_n \end{bmatrix} \right\| \le v_{1,n}, \tag{6.10c}$$

$$v_{2,n}\mathbf{I} - \mathbf{G}_n^{1/2}\mathbf{\Sigma}\mathbf{G}_n^{1/2} > \mathbf{0}, v_{2,n} \ge 0, \forall n \in \mathcal{N}, \tag{6.10d}$$

where $v_{1,n}$ and $v_{2,n}$ are slack variables.

Lastly, we relax $\mathbf{P}_4$ by dropping the constraint that $\mathbf{W}_k$ should have rank 1 for now, since it is not a convex one. The relaxed version of the problem is

$$\mathbf{P}_2 : \min_{\substack{\mathbf{W}_k, \mathbf{V}, \rho, \{v_{1i,k}\}, \{v_{2i,k}\}, \\ \{v_{1k,s}\}, \{v_{2k,s}\}, \{v_{1,n}\}, \{v_{2,n}\}}} \mathrm{Tr}\left(\sum_{k=1}^{K}\mathbf{W}_k + \mathbf{V}\right), \tag{6.11a}$$

s.t. (6.5), (6.7), (6.10), (6.2e), (6.2f), (6.2g). (6.11b)

Likewise, the coupling variables in (6.5b) and (6.7b) make $\mathbf{P}_2$ a non-convex problem. Thus we can still use the transformation in (5.17), which converts $\mathbf{P}_2$ into an equivalent optimization problem.

6.3 Maximum Harvested Energy Problem Formulation

In contrast to Section 6.2, where the minimum transmission power problem is considered, in the following we consider the optimization problem of maximizing the total harvested energy. This problem has important real-world applications, since most of the consumer electronics products are battery-driven and thus their energy efficiency is critical. In this section, we first formulate the problem, then we transform it in a convex way so that an existing software package can solve it efficiently. A one-dimensional search algorithm will be used. Furthermore, we also consider our previous pair of channel models.

In this section, we formulate the maximum harvested energy under the Gaussian CSI error model formulated is as follows:

$$\mathbf{P}_3 : \max_{\mathbf{W}_k \in \mathbb{C}^{M \times M}, \mathbf{V} \in \mathbb{C}^{M \times M}, \rho} \sum_{k=1}^{K} E_k^{\text{Practical}} \tag{6.12a}$$

s.t. (6.2b), (6.2e), (6.2e), (6.2f), (6.2g), (6.2h). (6.12b)

We first simplify the objective function and then a new approximation will be formulated based on the *Bernstein-type Inequality* [13, 194]. By involving a simple transformation, we arrive at:

$$\mathbf{P}_4 : \max_{\mathbf{W}_k, \mathbf{V}, \rho} \sum_{k=1}^{K} \mu_k \left\{ M_k - \epsilon_k \left(1 + \exp(-a_k(\tau_k - b_k)) \right) \right\} \tag{6.13a}$$

s.t. $\Pr(E_k^{\text{In}} \ge \tau_k) \ge 1 - \varpi, \ \forall \Delta\mathbf{h}_k \sim \mathcal{CN}(\mathbf{0}, \mathbf{H}_k), \forall k \in \mathcal{K},$ (6.13b)

(6.2b), (6.2e), (6.2e), (6.2f), (6.2g), (6.2h). (6.13c)

Observe however that the transformation from (6.12a) to (6.13a) and (6.13b) is not exactly equivalent. The equivalent form should let $E_k^{\text{In}} \geq \tau_k$ in (6.13b). However, by setting ϖ to be a very small value, the transformation can be valid and it is also consistent with our Gaussian CSI error model. By applying the *Bernstein-type Inequality I* [194], (6.13b) becomes,

$$\text{Tr}\left(\mathbf{H}_k^{1/2}\boldsymbol{\Sigma}\mathbf{H}_k^{1/2}\right) - \sqrt{-2\ln(\varpi)}v_{1k,s} + \ln(\varpi)v_{2k,s} + c_{k,s} \geq 0, \tag{6.14a}$$

$$c_{k,s} = \hat{\mathbf{h}}_k^\dagger\boldsymbol{\Sigma}\hat{\mathbf{h}}_k + \sigma_{k,S}^2 - \frac{\tau_k}{\rho}, \tag{6.14b}$$

$$\left\|\begin{bmatrix} \text{vec}\left(\mathbf{H}_k^{1/2}\boldsymbol{\Sigma}\mathbf{H}_k^{1/2}\right) \\ \sqrt{2}\mathbf{H}_k^{1/2}\boldsymbol{\Sigma}\hat{\mathbf{h}}_k \end{bmatrix}\right\| \leq v_{1k,s}, \tag{6.14c}$$

$$v_{2k,s}\mathbf{I} + \left(\mathbf{H}_k^{1/2}\boldsymbol{\Sigma}\mathbf{H}_k^{1/2}\right) \succ 0, v_{2k,s} \geq 0, \forall k \in \mathcal{K}, \tag{6.14d}$$

where $v_{1k,s}$ and $v_{2k,s}$, $k \in \mathcal{K}$ are slack variables.

We also relax the problem by dropping the constraint that the rank of $\mathbf{W}_k$ must be 1, and the optimization problem becomes

$$\mathbf{P}_5: \quad \max_{\substack{\mathbf{W}_k, \mathbf{V}, \rho, \{v_{1i,k}\} \\ \{v_{2i,k}\}, \{v_{1k,s}\} \\ \{v_{2k,s}\}, \{v_{1,n}\}, \{v_{2,n}\}}} \quad \sum_{k=1}^{K} \mu_k \left\{ M_k - \epsilon_k \left(1 + \exp(-a_k(\tau_k - b_k))\right)\right\},$$

$$\text{s.t. (6.14), (6.5), (6.10), (6.2e), (6.2f), (6.2g).} \tag{6.15a}$$

Still, the coupling variable in (6.14) can be tackled by fixing ρ. A similar one-dimensional search for ρ, together with a two-loop algorithm, can solve $\mathbf{P}_5$; the detailed step will be omitted here for space considerations.

6.3.1 Complexity Analysis

Similarly, under Gaussian error model, there are $3K(\frac{K+3}{2}) + 3N + 2$ linear constraints, $\frac{K(K+1)}{2} + 2K + N + 1$ LMI of size M, and $\frac{K(K+1)}{2} + K + N$ second-order cone (SoC) constraints. Thus, the complexity becomes:

$$C_{\text{com}}^G = \ln(\tau^{-1})n\sqrt{\Psi_{\text{comp}}^2}\left(\left(\frac{K(K+1)}{2} + 2K + N + 1\right)\right. \tag{6.16}$$

$$\left.[M^3 + nM^2] + 3K\left(\frac{K+3}{2}\right) + 3N + 2 + \left(\frac{K(K+1)}{2} + K + N\right)[(M^2 + M + 1)^2] + n^2\right)$$

where $\Psi_{\text{comp}}^2 = 3K^2 + 10K + 6N + 3$.

For the maximum harvested energy problem, with Gaussian error model, the complexity is $T'_{\text{max}}C_{\text{com}}^G$; correspondingly, T'_{max} is the number of unitary search.

6.4 Numerical Results

In this section, simulation results for characterizing the performance of the proposed robust beamforming were conceived with NOMA under both bounded and Gaussian estimation error models. Unless otherwise stated, the parameters are chosen as in Table 6.1.

Table 6.1 Simulation parameters

Parameters	Values
Number of SUs and PUs	$K = 3, N = 2$
Noise powers	$\sigma^2_{k,s} = 0.1, \sigma^2_p = 0.01$
Minimum required EH power	$P_{k,s} = 0.01$ Watt
Maximum tolerable interference of PUs	$P_{n,p} = -18$ dBm
Estimated channel gains	$\hat{\mathbf{h}}_k \sim \mathcal{CN}(\mathbf{0}, 0.8\mathbf{I})$ $\hat{\mathbf{g}}_n \sim \mathcal{CN}(\mathbf{0}, 0.1\mathbf{I})$
Outage probability threshold	$\xi_k = \xi_{k,s} = \xi_{n,p} = 0.05$
Gaussian CSI estimation	$\varpi^2_k = 0.001, \varpi^2_n = 0.0001$ [236]
Non-linear EH model	$M_k = 24$ mW, $a_k = 150$ $b_k = 0.014$ [24]

The performance comparison between bounded and Gaussian channel estimation models, as well as NOMA vs. OMA, is highlighted. To achieve a fair comparison between the two channel estimation error models, if the covariance matrices of the channel estimation error vector $\Delta\mathbf{h}_k$ and $\Delta\mathbf{g}_n$ under the Gaussian model are $\varpi^2_k\mathbf{I}$ and $\varpi^2_n\mathbf{I}$, respectively, then the bounded CSI radius under the worst-case scenario of φ_k and ψ_n should be [198]

$$\varphi_k = \sqrt{\frac{\varpi^2_k F^{-1}_{2M}(1 - \xi_k)}{2}}, \ \psi_n = \sqrt{\frac{\varpi^2_n F^{-1}_{2M}(1 - \xi_{n,p})}{2}}, \quad (6.17)$$

where $F^{-1}_{2M}(\cdot)$ represents the complementary cumulative distribution function (CCDF) of the chi-square distribution with $2M$ degrees of freedom.

6.4.1 Power Minimization Problem

Figure 6.1 shows the empirical CDFs of the minimum transmit power of the cognitive base station (CBS) for both the imperfect CSI estimation error models. The maximum power P_B is set to 2 Watts. For comparison, we also include the case of OMA, since it represents the traditional access technology. Observe that in order to reduce the inter-user interference, each OMA user relies exclusively on a single time slot. Thus, a total of K time slots are required instead of a single one in our scheme. To make a fair comparison, each SU's achievable data rate should be averaged over all K time slots, which becomes $R^{OMA}_k = \frac{1}{K}\log_2(1 + SINR^{OMA}_k)$. Reduced interference is achieved at the cost of a lower spectral and energy efficiency. We also observe that under both channel error models, the performance of NOMA is better than that of OMA. This is because for OMA, the lower spectral efficiency makes the SU data rate requirement harder to be satisfied. Hence the CBS has to apply a higher transmission power to compensate for that, which leads to a much higher energy consumption. Figure 6.1 is generated from 1,000 independent realizations of different channel conditions. As expected, the performance under perfect CSI is the best, since no additional power is used to compensate for the channel uncertainties. Furthermore, in both the OMA and NOMA schemes, the performance under the Gaussian CSI channel estimation is better than that under the bounded CSI channel estimations,

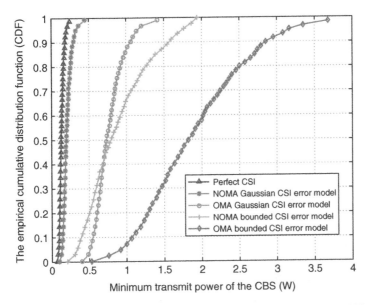

Figure 6.1 The empirical CDF of the minimum transmit power of the CBS under different channel conditions. CBS antenna number $M = 10$, $P_B = 2$ Watts, $R_{min} = 1$ bit/s/Hz.

as bounded CSI represents the worst-case scenario. Observe that the minimum power in the OMA-bounded CSI is over 2 Watts since we only limit the power of each time slot to 2 Watts and it is very likely that the total power over K slots will beyond that limit.

Figure 6.2 shows the minimum transmit power of the CBS as a function of the minimum required SNR of SUs, $\gamma_{k,min}$. As the SNR increases, the power increases under all CSI cases. Also, perfect CSI requires the least power, followed by NOMA relying on the Gaussian CSI error model, NOMA in the bounded CSI model, OMA Gaussian CSI model, and OMA-bounded CSI model. Besides, compared to OMA, the CBS power in NOMA grows more slowly. In the parameter setting, $\gamma_{k,min}$ plays a more important role in the constraints. For $\gamma_{k,min} = 2$ in the NOMA case, the equivalent SNR for OMA will be 26. Thus, the gap between OMA and NOMA further increases with the required SNR.

The impact of the CBS antenna number is illustrated in Figure 6.3a, where the performance with different CBS antenna numbers and channel uncertainties is plotted. Specifically, Figure 6.3a illustrates how the number of antennas affects the overall performance. The power required increases, when the SNR of SUs grows, regardless of how many antennas are mounted at the CBS. It is also observed that the minimum power required decreases when the number of antennas increases, since a larger number of antennas result in a higher degree of freedom (DoF). Besides, we also notice that the performance under the Gaussian error model is better than that under the bounded channel error case. In Figure 6.4b, the impact of channel uncertainties is illustrated. We set $\psi_n^2 = \varphi_k^2 = [0.01 : 0.05]$, the corresponding covariance matrices in Gaussian CSI estimation error scenario also change according to (6.17). Clearly, channel estimation error affects the bounded CSI scenario the most, since under worst-case CSI, the channel estimation error channel becomes worse; thus, it needs more power to meet the data rate constraints. Nevertheless, the channel

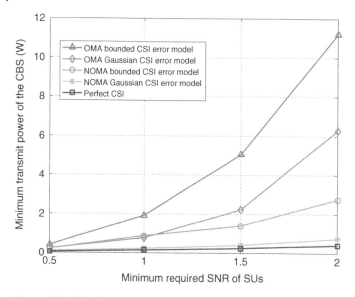

Figure 6.2 The minimum transmit power of the CBS vs. the required SNR of SUs for $M = 10$, $P_B = 8$ Watts.

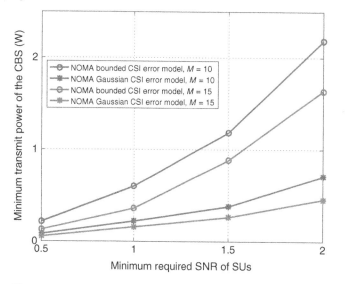

Figure 6.3 Impact of the number of CBS antennas on the minimum transmitted power required in two imperfect CSI scenarios, $M = 15$, $R_{min} = 1$ bit/s/Hz, $P_B = 8$ Watts.

estimation error does not have much impact on the Gaussian channel estimation error scenario.

6.4.2 Energy Harvesting Maximization Problem

In this section, we present simulation results where maximum EH is the objective function. The CBS power is $P_B = 2$ Watts, other parameters are the same as in Table 5.1. Figure 6.5

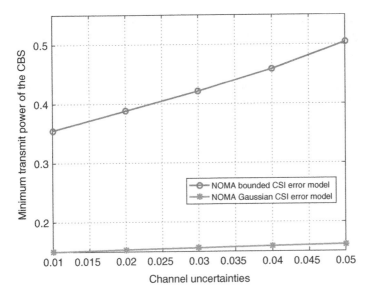

Figure 6.4 Impact of channel uncertainties ψ_n and φ_k on the overall minimum transmit power of the CBS, $M = 15$, $R_{min} = 1$ bit/s/Hz, $P_B = 8$ Watts.

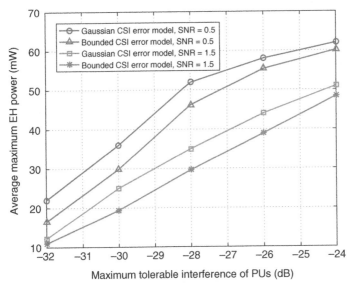

Figure 6.5 Average maximum EH power under different interferences tolerated by the PUs, $M = 10$.

characterizes the average maximum EH power vs. the interference tolerated by the to PUs. One can observe that the energy harvested monotonically increases, when the maximum interference tolerated by the PUs grows, where a higher $P_{n,p}$ allows for a larger transmission power, leading to the increase of the harvested energy. Additionally, we can see that under the Gaussian channel estimation error, the performance is better than that under the bounded channel estimation error case. When the channel conditions are better, less power is required for satisfying the data rate requirements. Hence more power can be reserved for EH. This also explains that when the required SNR is low, a high EH power can be achieved.

The impact of minimum SNRs required by the SUs is illustrated in Figure 6.6. The number of CBS antennas is $M = 10$ and the interference threshold $P_{n,p}$ is set to -24 dBm. We also

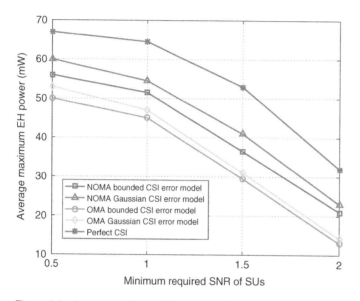

Figure 6.6 Average maximum EH power vs. the minimum SNR required by the SUs, $M = 10$.

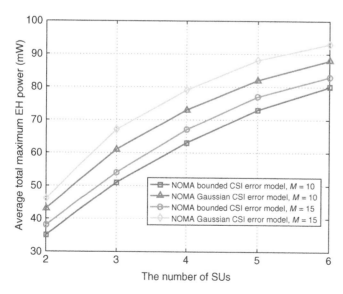

Figure 6.7 Average total EH power vs. the number of SUs for $P_{n,p} = -24$ dBm, $r_{min} = 1$ bit/s/Hz.

list the results for the OMA cases. As expected, the average maximum EH power decreases, when the required SNR increases. Similar observations show that under perfect CSI, the performance is the best, while the OMA-bounded CSI estimation scenario is the worst. Moreover, we can see that the maximum EH power decreases significantly when the SNR grows. This is because more power has to be used for information detection, which leaves less power for energy harvesting.

Figure 6.7 shows the average total EH power vs. the number of SUs. It can be observed that the total EH power grows, when the number of SUs increases, since more nodes participate in the harvesting process. Additionally, we can see that when the number of antennas is higher, more EH power can be achieved. This is because more antennas give a higher system DoF; therefore, less power is sufficient for information detection.

6.5 Summary

This chapter extends Chapter 5 and considered a more realistic Gaussian CSI estimation model. A similar min power and max EH problem is formulated followed by solutions and numerical results. As for future research directions, the system model can be generalized to account for more use cases, for example, considering the physical layer security and the interference arising from multiple cells.

7

Mobile Edge Computing in 5G Wireless Networks

7.1 Background

In Chapters 2–6, we mainly addressed the spectral and large number of device connection challenges for the future 5G systems. However, the ever-increasing demand for various applications such as gaming, autonomous driving, and augmented reality/virtual reality (AR/VR) have been recognized as one of the driving forces for the prosperity of the smart devices [142]. Due to the limitations on size, battery, and cost, these small-size smart devices can experience performance bottleneck when computation-intensive tasks need to be executed. One option is to deploy centralized services such as cloud centers to help the data processing. However, cloud servers can be located far away, which can inevitably cause longer end-to-end transmission delay [142].

In contrast to the centralized infrastructure, recent network paradigms such as MEC tend to allocate resources to devices in close proximity for joint processing. For example, the work in [181] used unmanned aerial vehicles (UAVs) to help D2D wireless networks [109]. This paradigm shift can effectively reduce the long backhaul latency and energy consumption, as well as support a more flexible infrastructure in a cost-effective way. Furthermore, MEC together with virtual machine (VM) migration can effectively increase the scalability [154] while reducing service delay [153]. Due to these advantages, MEC has attracted extensive research attention in various vertical segments.

One important feature of MEC is performing computation offloading, which leverages the powerful MEC servers in proximity and sends the computation-intensive tasks to MEC servers for processing. It can help overcome the physical limitations of local small devices. Current research involves two categories of offloading: binary [15] and partial [126–173, 178, 180–194, 196–220]. Binary offloading executes the task as a whole, either locally or in the MEC server, while partial offloading assumes the task can be partitioned into two parts, one for local processing and one for offloading. Even though the former is easier in implementation, for a very large dataset, partial offloading can help reduce the latency and energy consumption on the local devices more effectively.

Previous works either target on minimizing the total energy consumption or maximizing total computed bits. Energy-efficient communication has received tremendous industrial and academic attention in various systems such as multi-hop and heterogeneous networks [78]. By applying energy efficiency as the performance metric, QoS can be obtained, together with a reduction on energy consumption [209]. Energy efficiency defined in traditional communication systems in bits transmitted per Joule is an important metric to evaluate the overall system energy consumed. However, in the new communications systems, there exist a large number of computation-constrained and power-limited devices (such as IoT devices) that will need to support delay-critical yet computation-intensive tasks. Offloading through communications to MEC servers in order to compute the tasks timely becomes critically important to meet the short delay budget requirement, while communications throughput requirements may become secondary. To capture the efficiency of energy used for both computing and communication in such a scenario, we propose the metric computation efficiency, which is defined as the number of total computed bits divided by the energy consumed. We argue that this metric is more appropriate since it can measure how efficient the system is, in terms of computed bits per Joule, for a system involving massive computation needs.

Our work expands [199] and [209] in two major aspects. Firstly, we consider maximizing the computation efficiency instead of purely maximizing computed data bits or minimizing energy consumption compared with [199]. Secondly, we combine local computing and data offloading in a hybrid approach instead of offloading only [209]. This chapter presents the following advances in the MEC offloading system.

1. We propose a new performance metric in MEC networks: computation efficiency, which is defined as the number of computed bits divided by the corresponding energy consumption. Computation efficiency can drive toward efficient on-board power utilization while achieving satisfactory QoS.
2. The fundamental trade-off between local computing and data offloading is analyzed. Results show that with practical parameter settings, when data size is small, more data will be processed locally. But when the data grows, offloading will play a more important role in improving the computation efficiency.

7.2 System Model

In this chapter, we consider a downlink MEC network which consists of one MEC server and K randomly located UEs [173]. The server has a single antenna and so does each UE. Assume the channel between the server and the UE is a block-fading-based model, i.e. the channel remains constant during a time slot with length T but varies from time to time. The channel state information is assumed to be available at the server. At the beginning of a particular time T, each UE has a computation-intensive task to compute. Due to the computation resource limit or power limit or both, these tasks are offloaded to the nearby MEC server for a more powerful processing if needed. In this chapter, we assume the task-input bits are bit-wise independent and can be arbitrarily divided into different groups and executed by different entities in MEC system, e.g. parallel execution at the mobile and MEC server [199]. Partial offloading is used here. Thus the system can support data offloading and local computing simultaneously.

7.2.1 Data Offloading

Denote the set of UEs as $\mathcal{K} = \{1, 2, \ldots, K\}$. A UE can offload part of the computation bits to the server. To reduce the interference between different UEs, UEs doing offloading are allocated a portion of T and transmit sequentially, such as in the TDMA mode. Specifically, let g_k, p_k, and t_k, respectively, represent the channel between the server and UE k, the transmission power, and time duration allocated to UE k. The total number of offloaded bits is $r_k = B \log_2 \left(1 + \frac{p_k g_k}{\sigma^2}\right) t_k$, $\forall k \in \mathcal{K}$, where σ^2 is the noise power and B is the system bandwidth.

Under this mode, the corresponding energy consumption for UE k is $e_k = p_k t_k + p_r t_k$, where $p_k t_k$ denotes the over-the-air information transmission energy consumption, and p_r is the constant circuit power for transmit signal processing, which is the same for all UEs.

7.2.2 Local Computing

In addition to offloading, part of the bits can be computed locally by UEs. Let C_k be the number of computation cycles needed to process one bit of data for UE k. Clearly, each UE can compute the data throughout the entire block T. Furthermore, f_k denotes the processor's computing speed in the unit of cycles per second, and similar to [15], this speed holds constant. Therefore, the total number of bits locally computed is $r_k^{\text{local}} = \frac{T f_k}{C_k}$. The energy consumption of local computing is modeled as a function of the processor speed f_k. Specifically, $E_k^{\text{local}} = \epsilon_k f_k^3 T$, where ϵ_k is the computation energy efficiency coefficient of the processor's chip [15, 197].

7.3 Problem Formulation

In this section, we form an optimization problem that maximizes the total computation energy efficiency among all UEs. Mathematically, the problem is expressed as follows.

$$\mathbf{P}_1 \quad \max_{\{t_k\}, \{f_k\}, \{p_k\}} \sum_k w_k \frac{B \log_2 \left(1 + \frac{p_k g_k}{\sigma^2}\right) t_k + \frac{T f_k}{C_k}}{\epsilon_k f_k^3 T + p_k t_k + p_r t_k} \tag{7.1a}$$

$$\text{s.t.} \quad C1 : \sum_k t_k \leq T, \tag{7.1b}$$

$$C2 : B \log_2 \left(1 + \frac{p_k g_k}{\sigma^2}\right) t_k + \frac{T f_k}{C_k} \geq L_k, \forall k, \tag{7.1c}$$

$$C3 : \epsilon_k f_k^3 T + p_k t_k + p_r t_k \leq E_k^{th}, \forall k, \tag{7.1d}$$

$$C4 : 0 \leq f_k \leq f_k^{\max}, \forall k, \tag{7.1e}$$

$$C5 : t_k \geq 0, \forall k, \tag{7.1f}$$

where w_k is the weighting factor that can be used to prioritize different QoS requirements of UEs. $\mathbf{P}_1$ is a resource allocation problem that optimizes the offloading transmission time t_k and power p_k, as well as local computing chip frequency f_k. C_1 states that all the tasks should be completed before the end of the block. Notice that here we omit the processing and transmission time at the server by following [15, 199]. L_k in C_2 denotes the minimum data bits for computing for UE k. E_k^{th} in C_3 is the total energy available in UE k. C_4 defines the maximum CPU frequency of each UE.

The above problem is non-convex since the objective function involves sum-of-ratio maximization. Also, the coupling of some variables makes the optimization problem even more complicated. To address the coupling problem, let $P_k = p_k t_k$. Besides, for notational brevity, denote $R_k(P_k, t_k, f_k) = B \log_2 \left(1 + \frac{P_k g_k}{t_k \sigma^2}\right) t_k + \frac{T f_k}{C_k}$, and $E_k(P_k, t_k, f_k) = \epsilon_k f_k^3 T + P_k + p_r t_k$. We first employ simple transformations and the original problem becomes:

$$\mathbf{P_2} : \quad \max_{\{t_k\}, \{f_k\}, \{P_k\}, \{\beta_k\}} \sum_k w_k \beta_k \tag{7.2a}$$

s.t. $\quad C1 \ : \ R_k(P_k, t_k, f_k) \geq \beta_k E_k(P_k, t_k, f_k),$ \hfill (7.2b)

$$C2 \ : \ \sum_k t_k \leq T, \tag{7.2c}$$

$$C3 \ : \ t_k \geq 0, \forall k, \tag{7.2d}$$

$$C4 \ : \ 0 \leq f_k \leq f_k^{\max}, \forall k, \tag{7.2e}$$

$$C5 \ : \ R_k(P_k, t_k, f_k) \geq L_k, \tag{7.2f}$$

$$C6 \ : \ E_k(P_k, t_k, f_k) \leq E_k^{th}, \forall k. \tag{7.2g}$$

Lemma 7.1 *For $\forall k$, if $(\{t_k^*\}, \{f_k^*\}, \{P_k^*\}, \{\beta_k^*\})$ is the optimal solution of $\mathbf{P_2}$, there must exist $\{\lambda_k^*\}$ such that $(\{t_k^*\}, \{f_k^*\}, \{P_k^*\})$ satisfies the Karush–Kuhn–Tucker condition of the following problem for $\lambda_k = \lambda_k^*$ and $\beta_k = \beta_k^*$.*

$$\mathbf{P_3} : \quad \max_{\{t_k\}, \{f_k\}, \{P_k\}} \sum_k \lambda_k (w_k R_k - \beta_k E_k) \tag{7.3a}$$

s.t. (7.2c) – (7.2g). \hfill (7.3b)

Furthermore, $(\{t_k^*\}, \{f_k^*\}, \{P_k^*\})$ satisfies the following equations for $\lambda_k = \lambda_k^{\star}$ and $\beta_k = \beta_k^{\star}$:

$$\lambda_k = \frac{w_k}{E_k(P_k, t_k, f_k)}, \quad \beta_k = \frac{w_k R_k(P_k, t_k, f_k)}{E_k(P_k, t_k, f_k)}, \forall k. \tag{7.4}$$

Lemma 7.1 can be proved by taking the derivative of the Lagrange function of $\mathbf{P_2}$. λ_k is the non-negative multiplier of (7.2b). A detailed proof can be obtained in [92]. *Lemma 7.1* implies that the optimal solution of $\mathbf{P_2}$ can be obtained by solving the equations of (7.4) among the solutions of $\mathbf{P_3}$.

The Lagrange function of $\mathbf{P_3}$ is

$$\mathcal{L}(t_k, P_k, f_k, \alpha_k, \mu_k, \theta_k, n_k, m) \tag{7.5}$$
$$= \sum_k \lambda_k (w_k R_k - \beta_k E_k) - \sum_k \alpha_k (E_k - E_k^{th})$$
$$- \sum_k \mu_k (L_k - R_k) - \sum_k n_k (f_k - f_k^{\max}) - m \left(\sum_k t_k - T\right),$$

where $\alpha_k, \mu_k, \theta_k, n_k$, and m are non-negative Lagrange multipliers for the respective constraints. It can be readily proved that $\mathbf{P}_3$ is convex for given λ_k and $\beta_k, \forall k$, and satisfies Slater's condition. Thus, strong duality holds between the primal and dual problems, which means solving $\mathbf{P}_3$ is equivalent to solving the dual problem. Notice that the dual function is
$$\psi(\alpha_k, \mu_k, \theta_k, n_k, m) = \max_{\{t_k\}, \{f_k\}, \{P_k\}} \mathcal{L}\left(t_k, P_k, f_k, \alpha_k, \mu_k, \theta_k, n_k, m\right).$$ The dual problem becomes

$$\mathbf{P}_4 : \min_{\substack{\alpha_k, \mu_k, \\ \theta_k, n_k, m}} \psi(\alpha_k, \mu_k, \theta_k, n_k, m). \tag{7.6}$$

In the following, we first obtain the optimal solutions for the given auxiliary variables (λ_k, β_k) and Lagrange multipliers $(\alpha_k, \mu_k, \theta_k, n_k, m)$. Then the Lagrange multipliers are updated via gradient descent method. Lastly, the auxiliary variables are updated as well.

7.3.1 Update p_k, t_k, and f_k

Equation (7.5) can be re-organized as

$$\begin{aligned} &\mathcal{L}(t_k, P_k, f_k, \alpha_k, \mu_k, \theta_k, n_k, m) \\ &= \sum_k \left((\lambda_k w_k + \mu_k) R_k - (\alpha_k + \lambda_k \beta_k) E_k - n_k f_k - m t_k \right. \\ &\quad + \left. \alpha_k E_k^{th} - \mu_k L_k + n_k f_k^{max} \right) + mT. \end{aligned} \tag{7.7}$$

To maximize the dual function, $\psi(\alpha_k, \mu_k, \theta_k, n_k, m)$ can be decomposed into K sub-problems. Specifically, the k-th problem is

$$\begin{aligned} \psi_k &= \max_{\{t_k\}, \{f_k\}, \{P_k\}} \mathcal{L}_k(t_k, P_k, f_k, \alpha_k, \mu_k, \theta_k, n_k, m) \tag{7.8} \\ &= \max_{\{t_k\}, \{f_k\}, \{P_k\}} (\lambda_k w_k + \mu_k) R_k - (\alpha_k + \lambda_k \beta_k) E_k - n_k f_k - m t_k + \Psi, \end{aligned}$$

where Ψ denotes the constant value that is irrelevant to the optimizing variables.

Proposition 7.1 *The optimal transmit power and duration for the k-th UE should be*

$$p_k^* = \left[\frac{(\lambda_k w_k + \mu_k) B}{(\lambda_k \beta_k + \alpha_k) \ln 2} - \frac{\sigma^2}{g_k} \right]^+ \text{and} f_k^* = \sqrt{\left[\frac{\left(\frac{(\lambda_k w_k + \mu_k)}{c_k} - n_k \right) \frac{1}{c_k}}{3(\lambda_k \beta_k + \alpha_k)} \right]^+} \text{ respectively, where } [x]^+ = \max(x, 0).$$

Proof: Taking the derivative of the Lagrange function ψ_k w.r.t. P_k yields

$$\frac{\partial \psi_k}{\partial P_k} = \frac{(\lambda_k w_k + \mu_k) B t_k g_k}{(t_k \sigma^2 + P_k g_k) \ln 2} - \lambda_k \beta_k - \alpha_k. \tag{7.9}$$

Let $\frac{\partial \psi_k}{\partial P_k} = 0$, the optimal P_k^* can be obtained. Notice that the optimal p_k^* is equal to $\frac{P_k^*}{t_k}$. Similarly, let $\frac{\partial \psi_k}{\partial f_k} = 0$, we can get the optimal expression for f_k.

- *Remark*: In order to maximize EE, user k with a higher channel gain g_k should transmit with a higher power p_k. This can be seen from the optimal expression of p_k^*. Notice that the similar conclusion is also drawn in [209].

For t_k, the partial derivative expression of ψ_k w.r.t. t_k becomes

$$\frac{\partial \psi_k}{\partial t_k} = (\lambda_k w_k + \mu_k)B \log_2\left(1 + \frac{p_k g_k}{\sigma^2}\right) - (\alpha_k + \lambda_k \beta_k)(p_k + p_r) + m. \qquad (7.10)$$

Clearly, the optimization problem is a linear function of t_k. Therefore, the following problem can be solved efficiently by interior point methods.

$$\mathbf{P}_5 : \max_{\{t_k\}} \sum_k \lambda_k(w_k R_k - \beta_k E_k) \qquad (7.11a)$$

$$\text{s.t. } (7.2c), (7.2d), (7.2f), (7.2g). \qquad (7.11b)$$

7.3.2 Update Lagrange Multipliers

Now, we proceed to update the Lagrange multipliers α_k, μ_k, n_k, and m. From the problem definition, with known P_k, t_k, and f_k, the dual problem is always convex. Specifically, $\min_{\alpha_k, \mu_k, \theta_k, n_k, m} \psi(\alpha_k, \mu_k, \theta_k, n_k, m)$ is an affine function w.r.t. dual variables. Thus, we can apply the simple gradient method for the variable update. Specifically, we choose initial $\alpha_k(0)$, $\mu_k(0)$, $n_k(0)$, and $m(0)$ as the center of the ellipsoid which contains the optimal Lagrange variables. Then, we reduce the volume of the ellipsoid using gradient descent method as the following.

$$\alpha_k(i+1) = \alpha_k(i) + \Delta\alpha_k(E_k^* - E_k^{th}), \qquad (7.12a)$$

$$\mu_k(i+1) = \mu_k(i) + \Delta\mu_k(L_k - R_k^*), \qquad (7.12b)$$

$$n_k(i+1) = n_k(i) + \Delta n_k(f_k^* - f_k^{max}), \qquad (7.12c)$$

$$m(i+1) = m(i) + \Delta m\left(\sum_k t_k^* - T\right), \qquad (7.12d)$$

where $\Delta\alpha_k$, $\Delta\mu_k$, Δn_k, and Δm are the respective step size, i is the iteration index. Notice that all the Lagrange variables must be non-negative. If a negative value is obtained, the Lagrange variable will be set to 0 instead.

7.3.3 Update Auxiliary Variables

Lastly, the auxiliary variables λ_k and β_k are updated in the following way.

Notice that in *Lemma 7.1*, the optimal solution P_k^*, t_k^*, and f_k^* should also satisfy the following system conditions:

$$\beta_k E_k(P_k^*, t_k^*, f_k^*) - w_k R_k(P_k^*, t_k^*, f_k^*) = 0, \qquad (7.13)$$

$$\lambda_k E_k(P_k^*, t_k^*, f_k^*) - 1 = 0. \qquad (7.14)$$

Similarly, according to [92], we define functions for notational brevity. Specifically, let $T_j(\beta_j) = \beta_j E_k - w_k R_k$ and $T_{j+K}(\lambda_j) = \lambda_j E_k - 1, j \in \{1, 2, \ldots, K\}$. The optimal solution for λ_k

and β_k can be obtained by solving $\mathbf{T}(\lambda_k, \beta_k) = [T_1, T_2, \ldots, T_{2K}] = \mathbf{0}$. We can apply iterative method to update the auxiliary variables. Specifically,

$$\lambda_k(i+1) = (1 - \theta(i))\lambda_k(i) + \frac{\theta(i)}{E_k(P_k^*, t_k^*, f_k^*)}, \tag{7.15}$$

$$\beta_k(i+1) = (1 - \theta(i))\beta_k(i) + \theta(i)\frac{w_k R_k(P_k^*, t_k^*, f_k^*)}{E_k(P_k^*, t_k^*, f_k^*)}, \tag{7.16}$$

where $\theta(i)$ is the largest θ that satisfies $\|\mathbf{T}(\lambda_k(i) + \theta^l \mathbf{q}_{K+1:2K}^i, \beta_k(i) + \theta^l \mathbf{q}_{1:K}^i)\| \le (1 - z\theta^l)$ $\|T(\lambda_k(i), \beta_k(i))\|$, $\mathbf{q}$ is the Jacobian matrix of $\mathbf{T}$, $l \in \{1, 2, \ldots\}$, $\theta_l \in (0,1)$, and $z \in (0,1)$. Note that when $\theta(i) = 1$, it becomes the standard Newton method. To summarize, we list the detailed algorithm in **Algorithm 7.1**.

Algorithm 7.1 Computation Efficiency Maximization Algorithm

1: **Initialization:** the algorithm accuracy indicator t_1 and t_2, set $i = 0$, $\lambda_k(i)$ and $\beta_k(i)$
2: **while** $\|\mathbf{T}(\lambda_k, \beta_k)\| > t_1$ **do**
3:    **Initialization:** $\alpha_k(j)$, $\mu_k(j)$, $n_k(j)$, and $m(j)$, and let $j = 0$
4:    **while** $|\alpha_k(j+1) - \alpha_k(j)| > t_2$ **do**
5:       Calculate p_k^* and f_k^* based on *Proposition* 7.1.
6:       Solve for problem $\mathbf{P}_5$, obtain the timing variable t_k.
7:       Update Lagrange variables based on gradient descent method in (7.12).
8:       Let $j = j + 1$.
9:    **end while**
10:   Let $i = i + 1$, update auxiliary variables $\lambda_k(i+1)$ and $\beta_k(i+1)$ from (7.15) and (7.16).
11: **end while**
12: Output the optimal computation efficiency.

Notice that in the inner loop, the stop criterion can also be the convergence of other Lagrange multipliers or the condition that their combined value is less than a threshold.

7.3.4 Complexity Analysis

Since the algorithm involves the iteration process for three variables, we analyze the complexity in a sequential way. Firstly, p_k and f_k have a linear complexity with the user number K. The updating of Lagrange variables is of $\mathcal{O}(K^2)$ complexity since the total number of variables is $3K + 1$. Here $\mathcal{O}(x)$ means the upper bound for the complexity grows with order x. Finally, auxiliary variables λ_k and β_k have a complexity independent of K. Thus, our proposed algorithm has a total complexity in $\mathcal{O}(K^3)$.

7.4 Numerical Results

In this section, we present our simulation results of the joint offloading and computation scheme. The parameters are set as follows. The system bandwidth is $B = 200$ kHz, block length $T = 1s$, total number of UEs $K = 2$, $C_k = 10^3$ cycles needed for one bit raw data processing, the chip computing efficiency $\epsilon_k = 10^{-24}$, and static circuit power $p_r = 50$ mW. The channels between the MEC server and each UE are modeled as the joint effect of large-scale and small-scale fading, with $g_k/\sigma^2 = G_k h_k$, $G_1 = 7$, and $G_2 = 3$. h_k is the unitary Gaussian random variable. Lastly, the maximum computation capacity of each UE is set equally as $f_k^{max} = 10^9$ Hz. $E_1^{th} = E_2^{th} = 2$ Joule. All the results are averaged over different random channel realizations.

In Figure 7.1, we present the comparison results among three schemes, namely, the proposed scheme in this chapter, offloading only scheme, and local computing only scheme. We set $L_1 = L_2$ and $w_1 = w_2 = 1$, which means the minimum required data bits for all UEs are the same. In Figure 7.1, the computation efficiency of all the schemes decreases with the increase of the minimum required data bits. This suggests that the energy required to compute grows faster than the growth of the data bits. It is evident that our proposed algorithm outperforms other schemes. Additionally, we notice that when the data size is small, the proposed scheme's performance is closer to that of local computing only; when the data size grows, the performance will approach that of offloading only. This phenomenon can be explained by the following. Firstly, in the real-world applications, processor clock speed in a mobile device can reach MHz level. Thus, when the data size is relatively small, the preferred choice is to compute locally. Furthermore, based on the channel gain between UEs and the BS, and also the available bandwidth, data offloading may not be the ideal choice since it may take a longer time and a higher energy for small data offloading than for small data computing locally. On the other hand, when the data size is large, offloading to more computation powerful MEC server can become a much better choice. Moreover, the energy

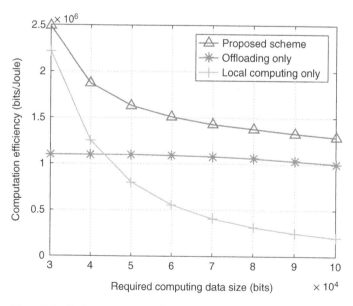

Figure 7.1 Performance comparison of different schemes.

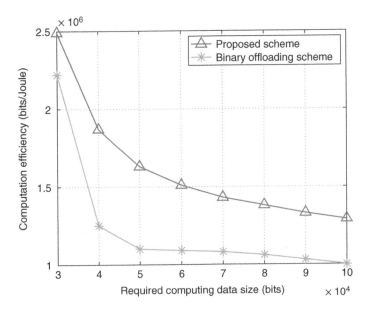

Figure 7.2 Performance comparison of our proposed scheme and the binary offloading.

decrease in local computing is more dramatic than the energy used in offloading when data size shrinks. According to the equation for local computing only, the computation efficiency is $\frac{r_k^{local}}{E_k^{local}} = \frac{1}{C_k \epsilon_k f_k^2} \propto \frac{1}{r_k^2}$, which indicates that its efficiency is inversely proportional to the square of the data size, while the offloading has a much slower deceasing rate thanks to the *log* function.

Compared with partial offloading, another MEC offloading scheme is the binary offloading, where each UE either completely offloads all the data to the MEC server or computes all the data locally. To compare its performance with our proposed scheme, we show the result in Figure 7.2. It can be seen that our proposed joint scheme outperforms the binary offloading in terms of computation efficiency, which indicates the superiority of the proposed algorithms.

Figure 7.3 illustrates the trade-off between two strategies: data offloading and local computing in our proposed scheme. The vertical axis represents the number of data bits (in percentage) calculated by either scheme with respect to the whole task. It can be readily shown that for both UEs, the local computing amount (in percentage) will decrease with the increase of the preset data amount. By contrast, data offloading plays a more and more important when the data becomes large. This can further prove our point in Figure 7.1, where the proposed scheme adaptively adjusts the amount of data that go through local computing or offloading. Additionally, for UE 1, the trade-off point happens around $L_1 = 4 \times 10^4$ and for UE 2 around $L_2 = 6 \times 10^4$. Since UE 1 has a better channel gain than UE 2, the influence from data offloading is more prominent; thus, the trade-off point is in an earlier position, while for UE 2, local computing continues to have a more influential role until the trade-off point $L_2 = 6 \times 10^4$.

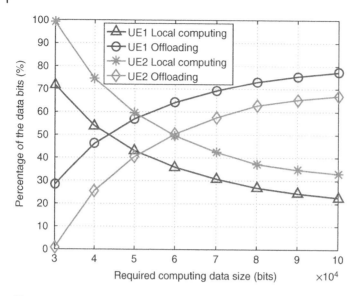

Figure 7.3 Trade-off between offloading and local computing.

7.5 Summary

In this chapter, we present a new evaluation metric in MEC systems, i.e. the computation efficiency. An optimization problem is formulated which aims at maximizing the total computation efficiency with weight factors. The problem is recognized as the sum-of-ratio problem and an iterative algorithm is applied in the outer loop. For the inner loop, the problem can be converted to standard convex optimization and, to gain a better insight, we propose to solve it via gradient descent method. Simulation results reveal the fundamental trade-off of two combined schemes: local computing and offloading.

8

Toward Green MEC Offloading with Security Enhancement

8.1 Background

In Chapter 7, our work proposed a new computation efficiency metric for evaluating MEC performance. Computation efficiency is similar to energy efficiency in wireless communication, which is defined as the number of bits computed divided by the total energy consumed. Several subsequent studies have adopted our metric and studied its performance under different scenarios. For example, Zhou and Hu [233] investigated computation efficiency maximization under wireless-powered MEC systems. It also compared two offloading schemes: joint and binary. An orthogonal frequency division multiple access (OFDMA)-based MEC system with computation efficiency maximization is proposed in [211], aiming for power and channel optimal allocation. Lastly, Zhang *et al.* [230] applied computation efficiency in an unmanned aerial vehicle (UAV)-enabled MEC offloading system, which also seeks for computation efficiency maximization.

On the other hand, wireless offloading will unavoidably suffer from potential malicious activities due to the existence of eavesdropper. From physical layer information-theoretic perspective, achievable data rate with eavesdropper can be modeled as the difference of mutual information from transmitter to receiver and from transmitter to eavesdropper, regardless of security protection mechanisms. It can be considered as the lower bound of the actual data rate. This simplified analysis model has been adopted in various studies [30, 208].

This chapter highlights the following research contributions.

1. We consider a secure offloading model, which allows wireless transmission with the presence of eavesdroppers. We adopt the physical layer security model from information-theoretic perspective and is irrelevant of encryption schemes.
2. Computation efficiency is applied as the main metric, which finds the balance between maximizing computation bits and minimizing total energy consumption.
3. An iterative algorithm together with convex approximation is proposed to tackle a non-convex problem and has good convergence speed and performance.

This chapter is organized as follows. Section 8.2 describes the secure offloading and computation model. An optimization problem is formulated in Section 8.3, followed by its approximation solution. Numeric results are presented in Section 8.4. Finally, Section 8.5 concludes this chapter.

5G and Beyond Wireless Communication Networks, First Edition. Haijian Sun, Rose Qingyang Hu, and Yi Qian.
© 2024 John Wiley & Sons Ltd. Published 2024 by John Wiley & Sons Ltd.

8.2 System Model

We consider a typical MEC system with one server and K UEs, the MEC system also mounts a wireless access point (AP) to communicate with other devices. We assume that both the AP and the UEs have a single antenna [177]. There also exists a malicious eavesdropper that tries to intercept confidential information. Denoted as *Eve*, the eavesdropper only has one antenna. At the beginning of a reference time t_s, each UE has a large number of computation-intensive tasks to be computed, because of the limited computation resources at UE, due to either device size or power constraints or both, UEs cannot finish their tasks before t_e. For timely processing, we require $t_e - t_s \leq T$. Hence, UEs will offload part of their computation bits to the MEC server, where a more powerful processing can be supported. In general, each UE supports the following operation modes (Figure 8.1).

8.2.1 Secure Offloading

In the presence of *Eve*, each UE must securely offload part of their task to the MEC server. We assume the channel between each UE and the AP in MEC server follows block static model where it remains the same within the block time T_1 ($T_1 \geq T$) but varies from one block to another. We denote the channel between UE k and the AP be h_k, where $h_k = l_k h_0$ is the joint effect of large-scale l_k and small-scale h_0 fading. Similarly, the channel between each UE and the *Eve* is g_k.

We consider the *active* eavesdropper scenario, where the *Eve* is also a user in the system, its listening and transmitting can be captured by UEs (from authentication, etc.). Therefore in this setting, we assume the channel between UE k and the *Eve* can be perfectly estimated, i.e. g_k can be perfectly estimated. Similar setting can also be found in [208].

Let the number of total bits be computed for each UE be L_k; since each UE cannot finish the calculation before the required time slot, it will send to MEC server for joint processing. Specifically, m_k is the number of bits that UEs offloads to the MEC securely. The signal received at the AP and *Eve* becomes:

$$y_k = h_k^H s_k + n_k, \forall k = 1, \dots, K, \tag{8.1}$$

$$y_e^k = g_k^H s_k + n_e^k, \forall k = 1, \dots, K, \tag{8.2}$$

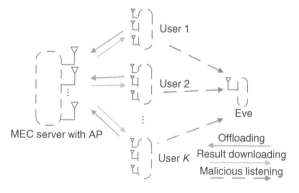

Figure 8.1 Secure MEC partial offloading model.

here, $s_k \in \mathbb{C}$ is the information-bearing signal for UE k, $n_k \in \mathcal{CN}(0, \sigma_k^2)$ and $n_e^k \in \mathcal{CN}(0, \sigma_{ek}^2)$ are the complex Gaussian noise at the AP and *Eve*, respectively. The secrecy rate, from information-theoretic perspective, is given as

$$R_{k,a}^{\text{sec}} = \left[\log\left(1 + \frac{p_k h_k^2}{\sigma_k^2}\right) - \log\left(1 + \frac{p_k g_k^2}{\sigma_{ek}^2}\right) \right]^+, \tag{8.3}$$

where $[a]^+ = \max(a, 0)$.

For offloading, the energy consumption consists of two parts: transmission and fixed circuit. In particular, $E_k^{\text{off}} = p_k t_k + p_r t_k$, where p_r, a constant, is the power of other circuit except for the transmission unit.

The rest $L_k - m_k$ bits will be calculated locally, which will be described below.

8.2.2 Local Computing

Traditionally, users will process all the computation locally. To model such a process, we first define some parameters. First, we assume user k's CPU needs C_k cycles to finish the computation of a single bit of data. Also, let f_k be the clock speed of the CPU. For simplicity, we assume the clock speed does not change. Each user is allowed to start the local computing from the beginning to the end of the process; thus, the total number of computation bits becomes $T f_k / C_k$.

Energy consumption for local computing can be modeled as $E_k^{\text{comp}} = \epsilon_k f_k^3 T$, where ϵ_k is the CPU energy coefficient [197, 209].

8.2.3 Receiving Computed Results

After receiving the computation task from each user, MEC server will start the calculation. When finished, it will send the result back to each user. Here, like [220, 233], we assume this process takes negligible time because of two reasons: 1) MEC server has powerful multi-thread processor, 2) compared with data bits to be computed, result takes way less space; hence, downlink transmission is almost instant.

The whole process is illustrated in Figure 8.2.

8.2.4 Computation Efficiency in MEC Systems

Like previously mentioned, we define computation efficiency as the total number of calculated bits divided by total energy consumed. Thus, $CE_k = \frac{B R_{k,a}^{\text{sec}} t_k + \frac{T f_k}{C_k}}{p_k t_k + p_r t_k + \epsilon_k f_k^3 T}$ [173], where B is the bandwidth for offloading.

Figure 8.2 Time sharing offloading scheduling.

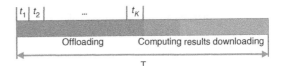

8.3 Computation Efficiency Maximization with Active Eavesdropper

We consider a green MEC system where the objective is to maximize the computation efficiency; the optimization problem is formulated as follows.

$$\mathbf{P_1} \quad \max_{\substack{\{t_k\},\{f_k\}, \\ \{m_k\},\{p_k\}}} \sum_k w_k \frac{BR^{sec}_{k,a} t_k + \frac{Tf_k}{C_k}}{\epsilon_k f_k^3 T + p_k t_k + p_r t_k} \tag{8.4a}$$

$$\text{s.t. } C1 \; : \; \sum_k t_k \leq T, \tag{8.4b}$$

$$C2 \; : \; BR^{sec}_{k,a} t_k \geq m_k, \forall k, \tag{8.4c}$$

$$C3 \; : \; L_k - \frac{Tf_k^{max}}{C_k} \leq m_k \leq L_k, \forall k, \tag{8.4d}$$

$$C4 \; : \; \epsilon_k f_k^3 T + p_k t_k + p_r t_k \leq E_k^{th}, \forall k, \tag{8.4e}$$

$$C6 \; : \; 0 \leq f_k \leq f_k^{max}, t_k \geq 0, p_k \geq 0 \; \forall k \tag{8.4f}$$

Our objective is to find the maximum value of the weighted summation for each UE's computation efficiency, w_k is the weight for UE k. The variables to be optimized here are the transmitted time for each UE t_k, the CPU frequency f_k, and the transmitted power for each user p_k. (8.4b) is the time constraint which requires the whole process ends before time T, (8.4c) combined with (8.4d) is the requirement for offloading and local computing rate. Furthermore, (8.4e) is the energy consumption constraint for each UE, where E_k^{th} is the maximum allowed energy. Lastly, f_k, t_k, and p_k should be non-negative variables, which is defined in (8.4f).[1]

Clearly, the formulated problem is non-convex, due to its sum-of-ratio objective function and C2, C4, C5, especially the coupling variables p_k and t_k. In the following, we tackle each non-convex term, mainly with successive convex approximation (SCA).

8.3.1 SCA-Based Optimization Algorithm

Firstly, we transform (8.4c) and (8.4e). Notice that we group these two constraints since they both involve with coupling variables. Let $\tilde{p}_k = p_k t_k$, then $R^{sec}_{k,a} t_k = t_k \log\left(1 + \frac{1}{\sigma_k^2}\tilde{p}_k h_k^2/t_k\right) - t_k \log\left(1 + \frac{1}{\sigma_{ek}^2}\tilde{p}_k g_k^2/t_k\right)$. For a function in the form of $f(x,y) = y \log\left(1 + \frac{x}{y}\right)$, it represents the entropy between x and y, and it is a concave function. Thus, $R^{sec}_{k,a}$ is still non-convex due to the difference of concave and convex functions. For approximation, we apply SCA

---

1 Theoretically, m_k should be an integer value, and (8.4d) becomes $m_k \in \left\{ L_k - \frac{Tf_k^{max}}{C_k}, L_k - \frac{Tf_k^{max}}{C_k} + 1, \ldots, L_k \right\}$. However, we simplify this and allow m_k to be fractional value, a feasible solution can take the round value if it is fractional, in [15], they also take the same approximation. When L_k is very large as the most real-world cases, the approximation has minimal impact to the original problem.

algorithm. Specifically, the entropy function has first-order Taylor series expansion at $(x, y) = (x_0, y_0)$

$$f(x,y) = y_0 \log\left(1 + \frac{x_0}{y_0}\right) + \left[\log\left(1 + \frac{x_0}{y_0}\right) - \frac{x_0}{x_0 + y_0}\right](y - y_0) + \frac{y_0}{x_0 + y_0}(x - x_0),$$

where (x_0, y_0) is the differentiable point. It is easy to verify that given $(x_0, y_0), f(x, y)$ becomes an affine form, which is convex. In optimization problems, we solve for (x, y) with given feasible point (x_0, y_0) first, then in the next round, (x_0, y_0) becomes the previous round's (x, y). The process will continue until converges. SCA-based approach works well in the iterative algorithms and received much attention recently.

In the following, to simplify notations, we first transform (8.4d) to equation sets below:

$$\tau_k \geq \frac{1}{\sigma_{ek}^2} \tilde{p}_k g_k^2, \tag{8.5a}$$

$$t_k \log\left(1 + \frac{N_k}{t_k}\right) - t_k \log\left(1 + \frac{\tau_k}{t_k}\right) \geq \frac{m_k}{B}, \tag{8.5b}$$

$$N_k \leq \frac{1}{\sigma_k^2} \tilde{p}_k h_k^2, \tag{8.5c}$$

where τ_k and N_k are auxiliary variables. It is easy to verify that the transformation is equivalent. Furthermore, (8.5b) should be transformed according to $f(\tau_k, t_k)$ and $f(N_k, t_k)$ that mentioned above.

8.3.2 Objective Function

Next, the objective function needs to be converted to the convex form as well. Currently, it is the summation of the fractional functions (represent each UE's computation efficiency). Traditional Dinkelbach's method cannot be applied directly since it can only deal with one fractional function. Instead, following [92], we can generalize Dinkelbach's algorithm to tackle one fractional function to multiple ones by a simple transformation.

$$\mathbf{P_3} \quad \max_{\{t_k\},\{f_k\},\{m_k\},\{\tilde{p}_k\},\{\beta_k\},\{\tau_k\},\{N_k\}} \sum_k w_k \beta_k \tag{8.6a}$$

s.t. $\quad C1 : R_k \geq \beta_k E_k, \forall k,$ \tag{8.6b}

$$C2 : \sum_k t_k \leq T, \tag{8.6c}$$

$$C3 : L_k - \frac{T f_k^{\max}}{C_k} \leq m_k \leq L_k, \forall k, \tag{8.6d}$$

$$C4 : \epsilon_k f_k^3 T + \tilde{p}_k + p_r t_k \leq E_k^{th}, \forall k, \tag{8.6e}$$

$$C5 : 0 \leq f_k \leq f_k^{\max}, t_k \geq 0, \tilde{p}_k \geq 0 \ \forall k, \tag{8.6f}$$

$$C6 : (8.5a)–(8.5c). \tag{8.6g}$$

Here, we let $R_k = BR_k^{\text{sec}} t_k + \frac{Tf_k}{C_k}$, and $E_k = \epsilon_k f_k^3 T + \tilde{p}_k + p_r t_k$, for notational simplicity. β_k is the auxiliary variable.

$$\max_{\substack{\{t_k\},\{f_k\},\{m_k\}, \\ \{\tilde{p}_k\},\{\tau_k\},\{N_k\}}} \sum_k \lambda_k \left\{ (v_k^{(i)} - \theta_k^{(i)})(t_k - t_k^{(i)}) + \frac{t_k^{(i)}}{t_k^{(i)} + N_k^{(i)}}(N_k - N_k^{(i)}) \right. \tag{8.7a}$$

$$\left. - \frac{t_k^{(i)}}{t_k^{(i)} + \tau_k^{(i)}}(\tau_k - \tau_k^{(i)}) \right\} w_k B + \sum_k \frac{\lambda_k w_k Tf_k}{C_k} - \sum_k \lambda_k \beta_k \left(\epsilon_k f_k^3 T + \tilde{p}_k + p_r t_k \right)$$

$$\text{s.t. } B \left\{ (v_k^{(i)} - \theta_k^{(i)})(t_k - t_k^{(i)}) + \frac{t_k^{(i)}}{t_k^{(i)} + N_k^{(i)}}(N_k - N_k^{(i)}) \right. \tag{8.7b}$$

$$\left. - \frac{t_k^{(i)}}{t_k^{(i)} + \tau_k^{(i)}}(\tau_k - \tau_k^{(i)}) + \varphi_k^{(i)} \right\} + \frac{Tf_k}{C_k} \geq L_k,$$

$$(8.6\text{c}), (8.6\text{e}), (8.6\text{f}), (8.5\text{a}), (8.5\text{c}) \tag{8.7c}$$

$\mathbf{P}_3$ can be solved with the following *Lemma*.

Lemma 8.1 *For $\forall k$, if $(\{t_k^*\}, \{f_k^*\}, \{m_k^*\}, \{\tilde{p}_k^*\}, \{\beta_k^*\}, \{\tau_k^*\}, \{N_k^*\})$ is the optimal solution of* $\mathbf{P}_3$*, there must exist $\{\lambda_k^*\}$ such that $(\{t_k^*\}, \{f_k^*\}, \{m_k^*\}, \{\tilde{p}_k^*\}, \{\tau_k^*\}, \{N_k^*\})$ satisfies the Karush–Kuhn–Tucker (KKT) condition of the following problem for $\lambda_k = \lambda_k^*$ and $\beta_k = \beta_k^*$.*

$$\mathbf{P}_4 : \quad \max_{\{t_k\},\{f_k\},\{P_k\}} \sum_k \lambda_k(w_k R_k - \beta_k E_k) \tag{8.8a}$$

$$\text{s.t. } (8.6\text{c})-(8.6\text{g}) \tag{8.8b}$$

Furthermore, $(\{t_k^*\}, \{f_k^*\}, \{m_k^*\}, \{\tilde{p}_k^*\}, \{\tau_k^*\}, \{N_k^*\})$ satisfies the following equations for $\lambda_k = \lambda_k^\star$ and $\beta_k = \beta_k^\star$:

$$\lambda_k = \frac{w_k}{E_k}, \quad \beta_k = \frac{w_k R_k}{E_k}, \forall k. \tag{8.9}$$

Please refer to [92] for the detailed proof.

Based on the above *Lemma*, $\mathbf{P}_3$ can be solved iteratively. At each iteration, the objective function becomes a convex one with giving λ_k and β_k in (8.8a), then the auxiliary value of λ_k and β_k will be updated according to Section 8.3.3.

8.3.3 Proposed Solution to $\mathbf{P}_4$ with given (λ_k, β_k)

To summarize, for given (λ_k, β_k), a complete version of $\mathbf{P}_4$ is illustrated at the top of next page. Where $\theta_k^{(i)} = \left[\log \left(1 + \frac{\tau_k^{(i)}}{t_k^{(i)}} \right) - \frac{\tau_k^{(i)}}{\tau_k^{(i)} + t_k^{(i)}} \right]$, $v_k^{(i)} = \left[\log \left(1 + \frac{N_k^{(i)}}{t_k^{(i)}} \right) - \frac{N_k^{(i)}}{N_k^{(i)} + t_k^{(i)}} \right]$, and $\varphi_k^{(i)} = t_k^{(i)} \log(1 + N_k^{(i)}/t_k^{(i)}) - t_k^{(i)} \log(1 + \tau_k^{(i)}/t_k^{(i)})$ is replaced for notational simplicity.

It is easy to verify that, for given $v_k^{(i)}, \theta_k^{(i)}, t_k^{(i)}, N_k^{(i)}$, and $\tau_k^{(i)}$ at each iteration, $\mathbf{P}_4$ is a convex optimization problem and can be solved by standard method such as interior-point algorithm. The optimized variable, once computed from the optimization problem, will be used to update $v_k^{(i)}, \theta_k^{(i)}, t_k^{(i)}, N_k^{(i)}$, and $\tau_k^{(i)}$ for the input of next iteration. Furthermore, the convergence of this iterative algorithm can be guaranteed by concave–convex procedure (CCCP).

Notice that for SCA algorithm, it is vital to select an appropriate initial value. The initial point used for iterative algorithm should be feasible for the optimization problem. We will discuss more details in the simulation part.

8.3.4 Update (λ_k, β_k)

In this part, we give descriptions for updating the auxiliary variables λ_k and β_k.

Notice that in *Lemma 1*, the optimal solution $\tilde{p}_k^*, t_k^*$, and f_k^* should also satisfy the following system conditions:

$$\beta_k E_k(\tilde{p}_k^*, t_k^*, f_k^*) - w_k R_k(\tilde{p}_k^*, t_k^*, f_k^*) = 0, \tag{8.10}$$

$$\lambda_k E_k(\tilde{p}_k^*, t_k^*, f_k^*) - w_k = 0. \tag{8.11}$$

Similarly, according to [92], we define functions for notational brevity. Specifically, let $T_j(\beta_j) = \beta_j E_k - w_k R_k$ and $T_{j+K}(\lambda_j) = \lambda_j E_k - 1, j \in \{1, 2, \dots, K\}$. The optimal solution for λ_k and β_k can be obtained by solving $\mathbf{T}(\lambda_k, \beta_k) = [T_1, T_2, \dots, T_{2K}] = \mathbf{0}$. We can apply iterative method to update the auxiliary variables. Specifically,

$$\lambda_k(i+1) = (1 - \theta(i))\lambda_k(i) + \frac{\theta(i)}{E_k(\tilde{p}_k^*, t_k^*, f_k^*)}, \tag{8.12}$$

$$\beta_k(i+1) = (1 - \theta(i))\beta_k(i) + \theta(i)\frac{w_k R_k(\tilde{p}_k^*, t_k^*, f_k^*)}{E_k(\tilde{p}_k^*, t_k^*, f_k^*)}, \tag{8.13}$$

where $\theta(i)$ is the largest θ that satisfies $\|\mathbf{T}(\lambda_k(i) + \theta^l \mathbf{q}_{K+1:2K}^i, \beta_k(i) + \theta^l \mathbf{q}_{1:K}^i)\| \le (1 - z\theta^l)\|\mathbf{T}(\lambda_k(i), \beta_k(i)\|$, $\mathbf{q}$ is the Jacobian matrix of $\mathbf{T}$, $l \in \{1, 2, \dots\}$, $\theta_l \in (0,1)$, and $z \in (0,1)$. This update is also available in our prior study [173]. To summarize, we list the detailed algorithm in **Algorithm 8.1**.

Algorithm 8.1 Secure Computation Efficiency Maximization Algorithm

---

1: **Initialization:** the algorithm accuracy indicator u_1 and u_2, set $i = 0$, give initial values for $v_k^{(i)}, \theta_k^{(i)}, t_k^{(i)}, N_k^{(i)}, \tau_k^{(i)}, \lambda_k^{(i)}$, and $\beta_k^{(i)}$.

2: **while** $\|\mathbf{T}(\lambda_k, \beta_k)\| > u_1$ **do**

3:     **while** $|\alpha_k(j+1) - \alpha_k(j)| > u_2$ **do**

4:         Solve for problem $\mathbf{P}_4$, obtain the intermediate optimal values $\{t_k\}, \{f_k\}, \{m_k\}, \{\tilde{p}_k\}, \{\tau_k\}$, and $\{N_k\}$.

5:         Let $j = j + 1$.

6:     **end while**

7:     Let $i = i + 1$, update auxiliary variables $\lambda_k(i+1)$ and $\beta_k(i+1)$ from (8.12) and (8.13).

8: **end while**

9: Output the optimal computation efficiency.

8.4 Numerical Results

In this section, simulation results from our proposed scheme and algorithm are presented. Parameters for the simulation are given as follows. We assume the bandwidth for the system is $B = 200\,KHz$, number of users $K = 2$, time threshold $T = 1\,s$. For local computation, the CPU needs $C_k = 1000$ operations to process one bit of data. In addition, the scaling factor for energy consumption $\epsilon_k = 1 \times 10^{-24}$, CPU for each user has a maximum frequency $f_k^{max} = 1 \times 10^9\,Hz$. For offloading, we assume the channel from user to the server to be $h_k^2/\sigma^2 = H_k h_0$, where h_0 is the normal Gaussian variable. Similarly, $g_k^2/\sigma^2 = G_k h_0$. The value of H_k and G_k will be given later. We apply no bias to both users hence set $w_1 = w_2 = 1$. Lastly, the maximum allowed energy consumption is $E_k = 1\,Joule$.

In Figure 8.3, we show the convergence performance of our iterative algorithm. Here, we set $H_1 = 7, H_2 = 5, G_1 = G_2 = 1$, and $L_1 = L_2 = \{50{,}000, 60{,}000\}$. As a typical case, we only present the result for optimal time allocation t_k here. It can be seen that our iterative algorithm has a good convergence speed; it only takes around 6 iterations to achieve the optimal value. In fact, we test with different initial points and manage to get the same performance. The other observation from Figure 8.3 is that, when the required computation bits becomes larger, the transmission time is also longer. Intuitively, this explains that more offloading bits are required.

In Figure 8.4, the computation efficiency under different *Eves* channels is presented. The x-axis is the number of required total computation bits. The first observation is that, under all scenarios, computation efficiency decreases with the increasing data size. If we break down the two parts for computation efficiency, we can easily see that pure local computing efficiency is square proportionally decreasing with the increasing of data size. Also, if the bit size is large, offloading part is also decreasing, due to, in part of the circuit power in the denominator.

Additionally, Figure 8.4 also shows the relationship between computation efficiency with security threads from *Eve*. Specifically, we set channels from *Eve* to users to be different.

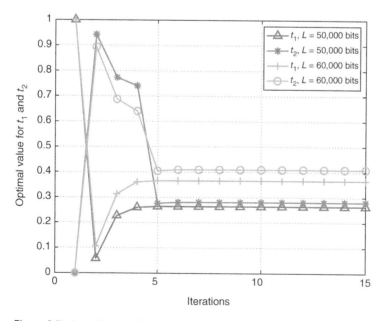

Figure 8.3 Iterative algorithm convergence analysis.

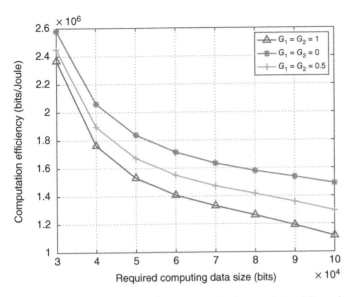

Figure 8.4 Computation efficiency vs. required computation bits under different *Eve* channels.

If the channel of *Eve* is stronger, we see a setback in the performance; this is due to the impact from offloading, where achievable data rate is smaller.

Lastly, we compare our proposed joint offloading and local computation scheme with other two schemes: local computing only and offloading only. Here, we set $G_1 = G_2 = 1$ (Figure 8.5). Clearly the proposed scheme outperforms both other two in terms of

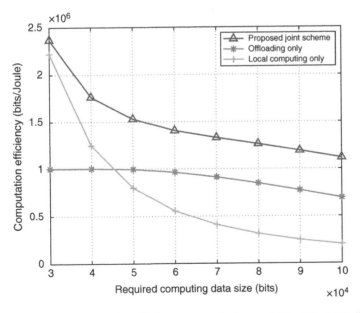

Figure 8.5 Computation efficiency vs. required computation bits under different computation scheme.

computation efficiency, which verifies the superiority of our scheme. Similar efficiency decreasing is also observed.

8.5 Summary

In this work, we studied the computation efficiency for a joint offloading and local computing scheme under possible *Eve*. We model the effect from *Eve* with physical layer security and mutual entropy. An optimization problem is formulated which considers some practical constraints. This non-convex problem is transformed with SCA and a general ratio iterative algorithm. In the future, we plan to generalize the chapter with multiple-antenna user/server and other access techniques.

9

Wireless Systems for Distributed Machine Learning

9.1 Background

Machine learning (ML) techniques especially deep learning have achieved remarkable performance in various applications such as object detection and content recommendation. State-of-the-art ML exploits the growing computation power of mobile devices that are capable of collecting, sharing, and processing data. Although such devices are still resource-constrained compared with the high-performance computing (HPC) centers, crowd-sourcing a large number of them can build powerful ML models.

Distributed ML requires devices to periodically share model attributes. Communication can become a severe challenge, especially when wirelessly connected devices participate in the distributed ML. Recently, federated learning (FL) is proposed to address this challenge by selecting a portion of the devices during each round for updating. Study shows FL can achieve multi-fold performance benefits including improving communication efficiency, preserving privacy, as well as handling heterogeneous datasets [128]. From wireless connection perspective, NOMA has been recognized as a new access technology in 5G to improve spectrum efficiency. NOMA allows multiple users to share the same radio resource simultaneously. To mitigate the interference, SIC is applied at the receiver side. SIC starts the decoding for the signal with the strongest received power and subtracts the decoded signal from the composite received message [229]. The process is sequentially carried out until the intended signals are all decoded. To further realize more efficient communication in FL, study in [106] found that 99.9% of the attributes exchanges are redundant and deep gradient compression was applied to reduce the message size for transmission. Li *et al.* [103] proposed a modification to tackle FL in heterogeneous networks. The above two works did not consider the constraint of the actual wireless communication process. The work presented in [6] exploited the medium access control (MAC) property and used both analog and digital communications to directly get the model average. However, they did not consider the effects from wireless fading channels. Besides, gradient projection transmitted over MAC directly leads to a higher bit error rate (BER). In this chapter we propose to use NOMA in the uplink FL communication by considering the fading channels and adaptively compresses the gradient for the optimal transmission. This chapter highlights the following research contributions.

5G and Beyond Wireless Communication Networks, First Edition. Haijian Sun, Rose Qingyang Hu, and Yi Qian.
© 2024 John Wiley & Sons Ltd. Published 2024 by John Wiley & Sons Ltd.

- We utilize NOMA-enabled adaptive FL model update to reduce the aggregation latency. The study shows that NOMA can achieve superior performance in terms of communication latency compared with the traditional time division multiplexing access (TDMA) approach, which is commonly utilized in the existing FL schemes.
- Considering the capacity limitation in wireless fading channels, we further propose to apply adaptive quantization and sparsification to compress model updates in the uplink NOMA-based FL in order to save bandwidth. We demonstrate the effectiveness of the proposed scheme with several distinct datasets. Results show that the communication latency for the NOMA-based FL update with gradient compression is significantly reduced by at least 7× with no compromise on the test learning accuracy.

This chapter is organized as follows. Section 9.2 introduces the system model, FL update mechanism, and two compression techniques. Section 9.3 presents the NOMA transmission scheme, user scheduling, and adaptive transmission scheme. Simulation results are shown in Section 9.4 to verify the proposed schemes. Lastly, Section 9.5 concludes the chapter.

9.2 System Model

In this section, we introduce the FL model update and FL model compression schemes including quantization and sparsification. The main notations used in the chapter are summarized in Table 9.1.

9.2.1 FL Model Update

The system considers a total of N edge devices that distributively and collaboratively build a global learning model [178]. Each device or user collects and maintains its own raw data

Table 9.1 Summary of notations.

Notation	Definition
N; K; C	Number of edge devices connected to PS; number of edge devices participating FL in each round; fraction of participating devices, i.e. $C = K/N, 0 < C < 1$
$\mathbf{x}_j; \mathbf{y}_j; \theta_j$	Features of data point j; label of data point j; parameter set describe the mapping from $\mathbf{x}_j$ to $\mathbf{y}_j$
$f(\cdot); \eta$	Loss function; learning rate
$D_k; |D_k|$	Dataset on user k; cardinality of the dataset D_k
$h_k; L_k; h_0$	Channel coefficient of user k; large-scale fading of user k; small-scale fading, $h_0 \sim \mathcal{CN}(0,1)$
$p_k; \delta_k; d_k$	Power of user k; transmitter and receiver antenna gain; distance between user k and PS
$\lambda; \alpha; n_t$	Signal wavelength; path-loss exponent; additive noise
$\mathbf{g}_k; \mathbf{s}_k^t; R_k$	Gradient of user k; encoded gradient update from user k at communication round t; data rate of user k
$GS(\cdot); GQ(\cdot)$	Gradient sparsification function; gradient quantization function

locally. ML generally finds the mapping between features $\mathbf{x}_j$ and label $\mathbf{y}_j$. A loss function $f(\mathbf{x}_j, \mathbf{y}_j; \theta_j)$ is used to capture the error between this mapping. Typical loss functions used include linear regression and root-mean-squared error.

Each user performs learning algorithms locally. Essentially, local learning at user n aims to solve the following problem:

$$\min_\theta F_n(\theta) = \frac{1}{|\mathcal{D}_n|} \sum_{j \in \mathcal{D}_n} f(\mathbf{x}_j, \mathbf{y}_j; \theta_j). \tag{9.1}$$

Different from the traditional learning process where $\{\mathcal{D}_1, \mathcal{D}_2, \dots, \mathcal{D}_N\}$ are placed in the same location, FL can rely on the distributed stochastic gradient descent (SGD) method to perform update in each iteration. Specifically, the loss function in (9.1) can be generalized across multiple devices as:

$$\min_\theta f(\theta) = \sum_{n=1}^{N} \frac{|\mathcal{D}_n|}{|\mathcal{D}|} F_n(\theta), \tag{9.2}$$

where $|\mathcal{D}| = \sum_{n=1}^{N} |\mathcal{D}_n|$.

In each training round, FL selects a portion of the total devices to participate the global update. Initially the PS sets the model as θ^0 and sends it to all the users. Afterwards each user iteratively performs the local training and updates the gradient $\mathbf{g}_k = \nabla F_k(\theta)$. In the FL setting, each user can actually run multiple iterations on gradient calculations in each round. More specifically, in round t, user k calculates $\theta_k^t = \theta_k^t - \eta \nabla F_k(\theta)$ multiple times. The participating users then send their locally trained gradients to the PS for aggregation. The PS further calculates $\theta^{t+1} = \theta^t - \sum_{k=1}^{K} \frac{|\mathcal{D}_k|}{|\mathcal{D}|} \theta_k^t$ and sends θ^{t+1} to all the users for the next round update. A brief illustration of the FL model update is shown in Figure 9.1a.

Furthermore, each scheduled device needs to adjust their update size in order to fit that into the data rate supported by the dynamic fading channel. If user k's total update size exceeds the maximally allowable data rate m_k, model compression needs be applied. In the following we briefly introduce two lossy compression algorithms and their rationales.

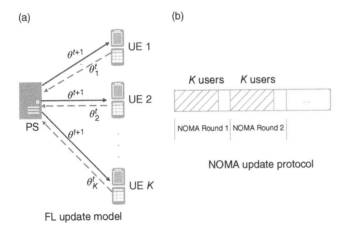

(a)

(b)

FL update model

Figure 9.1 An illustration of the proposed scheme. (a) A general FL model update. (b) NOMA update protocol in each round. Shaded area is for uplink and blank is for downlink.

9.2.2 Gradient Quantization

It is well-known that quantization can help compress the size of a large data. Standard algorithms in ML typically use 32-bit floating-point to represent each model parameter. Cost to store, transmit, and manipulate those data tends to be high. Alternatively, a simple implementation uses less bits for such a representation. Even though quantization creates "rounding errors," existing works show that this approach demonstrates a good model convergence [106] at the cost of more communication rounds. In this work, we adopt DoReFa scheme [235] suitable for quantizing gradients within $[-1, 1]$, which is true for most ML models. Specifically, mapping can be established with the function $q_k(x) = \frac{1}{a} \lfloor ax \rfloor$. Here, $\lfloor \cdot \rfloor$ rounds the value to the nearest integer, x is the gradient value, and $a = 2^b - 1$, where b is the quantization bit length.

9.2.3 Gradient Sparsification

Sparsification refers to the approach that sends the selected gradients instead of sending all of them. Empirical experiments have shown that a large portion of the gradient updates in a distributed SGD are redundant. Therefore, we can first map the smallest gradients to zero and then make a sparse update. For example, the threshold-based sparsification only keeps the gradients larger than a known threshold and sets the rest of small gradients to 0. The selection normally is based on the absolute gradient value. After sparsification, the non-zero gradients are uploaded to the PS. This chapter employs a similar method to perform gradient sparsification. Since the PS averages the gradient values in an element-wise way, it is important to take the non-zero index into consideration. Additionally, we update the relative distance (delta) between adjacent non-zero values rather than recording its absolute position. Moreover, the well-known non-linear coding called Golomb code is applied to encode the delta value, which uses variant-length bits to further save space. Specifically, for user k, the average number of bits used for encoding delta with Golomb coding is $\overline{b}_{pos}^k = b_k^* + \frac{1}{1-(1-r_s^k)^{2^{b_k^*}}}$, where $b_k^* = 1 + \lfloor \log_2(\frac{\log_2(\phi-1)}{\log_2(1-r_s^k)}) \rfloor$, $\phi = \frac{\sqrt{5}+1}{2}$, and r_s^k is the sparsification ratio.

9.3 FL Model Update with Adaptive NOMA Transmission

9.3.1 Uplink NOMA Transmission

In a typical wireless setting, traditional FL update uses TDMA for uplink transmission. The PS needs to wait until receiving the last user's message and then averages the received information from all the users. NOMA allows multiple users to share the uplink channel simultaneously. The channel coefficient between user k and the PS is $h_k = L_k h_0$. To simplify the analysis, we assume L_k follows the free-space path loss model $L_k = \frac{\sqrt{\delta_k}\lambda}{4\pi d_k^{\alpha/2}}$.

Let the encoded gradient update s_k^t be the transformation from θ_k^t in the local update stage. Additionally, we normalize the transmitted symbols $\|s_k^t\|_2^2 = 1$. According to NOMA principle, all the selected K users share the same bandwidth simultaneously. In particular, all the transmitted signals from multiple NOMA users are superposed [44]. The received signal at the PS at t thus can be expressed as:

$$y^t = \sum_{k=1}^{K} \sqrt{p_k} h_k s_k^t + n^t, \tag{9.3}$$

where s_k^t is symbolized representation of s_k^t at time slot t, $n^t \sim \mathcal{CN}(0, \sigma^2)$, σ^2 is noise variance of the received signal.

SIC is carried out at the PS side. Specifically, PS decodes the strongest signal first by treating others as interference. After successful decoding, PS subtracts the decoded signal from the superposed signal. The process stops until the PS decodes all the participants' messages. Without loss of generality, we assume $p_1 h_1^2 > p_2 h_2^2 > \cdots > p_K h_K^2$. Therefore, the achievable data rate for user k is:

$$R_k = \log_2 \left\{ 1 + \frac{p_k h_k^2}{\tau(\sum_{j=k+1}^{K} p_j h_j^2 + \sigma^2)} \right\}, \forall k = \{1, \ldots K-1\}, \tag{9.4}$$

where $\tau > 1$ accounts for performance degradation from finite length symbol, imperfect channel estimation, and decoding error, etc. User K is the last decoded user hence its rate is $R_K = \log_2(1 + \frac{p_K h_K^2}{\tau \sigma^2})$.

At the beginning of each round t, the PS notifies participating users to start the simultaneous transmission. The maximum number of allowable bits for user k is $m_k = BR_k t_k$, where B is the system bandwidth, t_k is the NOMA transmission duration. The proposed uplink NOMA transmission is shown in Figure 9.1b.

9.3.2 NOMA Scheduling

To select K users from a total of N to participate the model update, we should consider both the learning process and the communication process. The selection criterion is decided based on two rules: (1) NOMA fairness; (2) time budget.

1. NOMA fairness: During the PS update, the weighted average is applied, i.e. $\theta^{t+1} = \theta^t - \sum_{k=1}^{K} \frac{|D_k|}{|D|} \theta_k^t$. Therefore we use the following "effective update capacity"

$$R_{ef}^k = \frac{BR_k t_k}{|D_k|}, \tag{9.5}$$

to account for the actual contribution for the weighted average update. As discussed later on in Section 9.4, FL experiences performance degradation when data rate is heterogeneously distributed. Therefore, to make sure every user has a quality update, we use a widely accepted Jain's fairness index, which is defined as:

$$J_u = \frac{(\frac{1}{K} \sum_{k=1}^{K} R_{ef}^k)^2}{\frac{1}{K} \sum_{k=1}^{K} (R_{ef}^k)^2}. \tag{9.6}$$

For the maximum fairness, J_u should be close to 1. In practice, PS selects users with high effective update capacities and ensures J_u to be close to 1. We adopt a similar scheduling algorithm in [3].

2. *Time budget:* Another factor is the computation time at each device. Since NOMA is a synchronous system, the PS needs to set a hard time budget for the local computation. All the devices may have heterogeneous capacities; hence, they need to estimate the time spent on the training in each iteration. The scheduling used in this chapter selects those who can not only finish the calculation on time but also complete the most iterations of computing in each round.

9.3.3 Adaptive Transmission

For the selected user k in each round, we calculate the maximum throughput m_k under the NOMA scheme. The total bit length of gradient G is known once we have determined the ML structure. Thus the compression rate for quantization r_q^k is calculated as $r_q^k = \max\{\frac{G}{m_k}, 1\}$. The quantization bit length b_q^k is calculated by $b_q^k = \lfloor \frac{1}{r_q^k} 32 \rfloor$, where $\lfloor \cdot \rfloor$ takes the floor operation, 32 is bit-length of each parameter representation. Afterwards, every gradient value in user k is represented by bits with a length of b_q^k.

Similarly, under sparsification, compression ratio r_s^k can be calculated by solving the following equation

$$Gr_s^k + \frac{Gr_s^k - k}{32} b_{pos} = m_k. \tag{9.7}$$

Hence $r_s^k = \min\{r_s^k(n), 1\}$, $r_s^k(n)$ is the numeric solution for (9.7). Once r_s^k is obtained, we set $(1 - r_s^k)$ portion of the smallest gradient values to zero.

For both compression methods, it is important to keep the gradient residuals for the next round.[1] $\Delta\theta_k^t = \theta_k^t - GS(\theta_k^t)$ or $\Delta\theta_k^t = \theta_k^t - GQ(\theta_k^t)$, where $GS(\cdot)$ and $GQ(\cdot)$ are sparsification and quantization functions, respectively. By using them we can effectively reduce the compression accumulation errors. **Algorithm 9.1** summarizes the proposed scheme.

Algorithm 9.1 Adaptive FL Update with Uplink NOMA and Gradient Compression

---

1: **Initialization:** PS gives initial θ^0, maximum rounds T.
2: **for** each FL update round t **do**
3:      PS selects $K = CN$ users and calculates their maximum achievable data rates m_k. Then sends synchronous pilots, m_k, and θ^t to users.
4:      **for** each selected user k in parallel **do**
5:          Update local gradient one or multiple times: $\theta_k^t = \theta_k^t - \eta\nabla F_k(\theta)$, according to time budget.
6:          Based on m_k and size of gradient, apply either sparsification $GS(\theta_k^t)$ or quantization $GQ(\theta_k^t)$.
7:          Gradient residual $\Delta\theta_k^t$ will be kept locally for next round update.
8:          Send gradients to the PS at the beginning of synchronous time slot.
9:      **end for**
10:     PS applies SIC to decode gradient from K users.
11:     PS performs weighted average: $\theta^{t+1} = \theta^t - \sum_{k=1}^{K} \frac{|D_k|}{|D|} \theta_k^t$.
12: **end for**

---

1 Intuitively, some residuals can be accumulated to exceed the minimum threshold and can be used to help speed up the model convergence.

Note: Compared with the TDMA-based scheme, our proposed NOMA-based scheme achieves a lower latency performance. However, at the PS side, the signal decoding complexity is elevated as NOMA requires a multi-user detection (MUD) receiver to decode user messages successively.

9.4 Scheduling and Power Optimization

In Section 9.3, each round chooses K users randomly, and their transmission power is fixed. A careful user scheduling and power allocation can further improve system performance. In the following, we provide problem formulation and its solution [122].

9.4.1 Problem Formulation

Here we provide the formulated optimization problem with the following three constraints considered in our system model.

- $C1$: Each device can be scheduled at most once across different rounds.
- $C2$: At most K devices are allowed to participate the FL update in each round under NOMA.
- $C3$: Transmission power of each device in each round is bounded by a maximum value.

We aim to maximize a weighted sum rate of all participated devices, the optimization problem is formulated as

$$\max \sum_{m,t} w_m^t \Lambda_m^t R_m^t \tag{9.8a}$$

$$\text{s.t. } \sum_t \Lambda_m^t \leq 1, \forall m, \tag{9.8b}$$

$$\sum_m \Lambda_m^t \leq K, \forall t, \tag{9.8c}$$

$$0 \leq p_m^t \leq p_m^{t\,\max}, \forall (m,t) \in \mathcal{M} \times \mathcal{T}, \tag{9.8d}$$

$$\Lambda_m^t \in \{0,1\}, \forall (m,t) \in \mathcal{M} \times \mathcal{T}, \tag{9.8e}$$

where w_m^t is the data rate weight of device m scheduled at round t. In FL, PS performs weighted average to generate the current global model; hence, a natural selection for the data rate weight can be $w_m^t = \frac{|D_m|}{|D|}$, which also clearly outlines the significance of each device's update. $\Lambda_m^t = \{0,1\}$ is a binary variable that equals 1 if device m is scheduled at t and is 0 otherwise. Here, the constraint in (9.8b) corresponds to constraint $C1$, constraint in (9.8c) corresponds to constraint $C2$, and constraint in (9.8d) corresponds to constraint $C3$. Finding the maximum weight sum data rate under these constraints involves traversing all possible scheduling patterns, which possess very high complexity when the number of total devices is large and selected devices for scheduling are small, i.e. $M \gg K$. Toward that, we propose the following scheduling algorithm to address this complexity issue and power allocation to solve the optimization problem (9.8a).

9.5 Scheduling Algorithm and Power Allocation

Figure 9.2 shows the diagram of the user scheduling. Each column represents a FL round for model update, and there are a total of T columns. Each block in a specific column represents a scheduled user and at most K users are scheduled to participate FL update in each round. The power of the scheduled user k in round t is p_k^t. $(i_1, i_2, \ldots, i_K)$, $(j_1, j_2, \ldots, j_K)$ and $(l_1, l_2, \ldots, l_K)$ are different user combinations.

For the proposed joint scheduling and power allocation scheme, first, all possible user schedules are found. Then optimal power allocation is applied for each schedule to find the optimal one. The scheduling problem which aims to maximize weighted sum rate is transformed under graph theory. Specifically, we introduce the maximum weight independent set problem first. An independent set is a sub-graph of an undirected graph where there exists no edge between any two vertices. When the weight of each vertex is set to be equal to the sum data rate of users scheduled in the specific round, the sum of the weight of all vertices in an independent set equals to the sum data rate of a possible user schedule. The maximum weight independent set then corresponds to the schedule pattern that maximizes the sum data rate. The maximum weight independent set problem involves searching for all possible independent sets and then finding the maximum weight one. Thus a critical step is to construct the scheduling graph in order to find all the scheduling patterns.

9.5.1 Scheduling Graph Construction

Let S be the set that includes all the possible scheduling patterns for all the devices and rounds. $s \in S$ is a possible schedule. The scheduling graph can be constructed as follows. First, we need to generate vertices. In this graph, a vertex $v_j = (j_1, j_2, \ldots, j_K)t$ indicates that devices $j_1, j_2, \ldots, j_K$ are scheduled at time t. There are a total of $\binom{M}{K} \times T$ vertices. When creating the edges, the following constraints need to be satisfied.

- $C1$: Each device can be scheduled at most once.
- $C2$: At most K devices can be scheduled in one round.

For two vertices $v_i = (i_1, i_2, \ldots, i_K)t_i$ and $v_j = (j_1, j_2, \ldots, j_K)t_j$, if $i_k \in \{j_1, j_2, \ldots, j_K\}$, $\forall k = \{1, \ldots K\}$ (violates $C1$) or $t_i = t_j$ (violates $C2$), v_i and v_j are connected and an edge exists between these two vertices. Then when we select vertices from independent set,

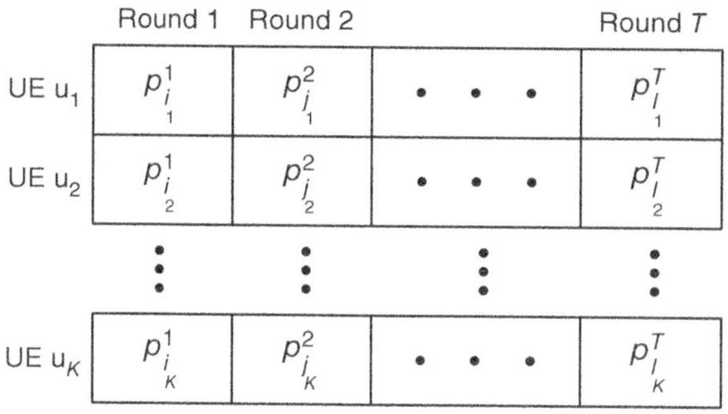

Figure 9.2 Scheduling diagram.

Figure 9.3 A scheduling graph example.

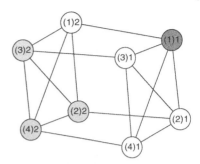

both C1 and C2 will be satisfied. Let us construct a scheduling graph example with $M = 4$, $K = 1$, and $T = 2$, as shown in Figure 9.3. In this case there are $\binom{4}{1} \times 2 = 8$ vertices. From this figure, we can find out that the possible independent sets for vertex $(1)1$ (green node) are $\{\{(1)1, (2)2\}, \{(1)1, (3)2\}, \{(1)1, (4)2\}\}$. Similarly, we can find all the independent sets for each vertex in the graph. Because of the edge connection constraints, each independent set has at most T vertices. Since the FL rounds are continuous and the number of FL rounds is T, the independent sets with T vertices are only considered.

9.5.2 Optimal scheduling Pattern

When scheduling graph is constructed, we calculate the weight of each vertex as sum data rate of users scheduled in a specified round, that is

$$w(v_j) = \sum_{k \in v_j} w_k^t R_k^t, \forall t \in s. \tag{9.9}$$

Then the sum of the weight of all vertices in an independent set equals the sum data rate of a possible schedule, that is

$$\sum_j w(v_j) = \sum_{k,t} w_k^t R_k^t, \forall (k,t) \in S. \tag{9.10}$$

where v_j represents vertex in an independent set.

The objective function in (9.8a) is actually equal to the problem maximizing the (9.10), which is the maximum weight independent set problem. The maximum weight sum rate problem then can be transformed as a maximum weight independent set problem. And the optimal schedule can be selected in the **Algorithm 9.2**:

Algorithm 9.2 Optimal scheduling selection.

1: **Require:** $\mathcal{M}, \mathcal{K}, \mathcal{T}, p_m^t$, and h_m^t.
2: Initialize $\mathbf{O} = \emptyset$
3: Construct scheduling graph G
4: Compute $w(v), \forall v \in G \; G \neq \emptyset$
5: $Q = \left\{ v | w(v) \geq \sum_{u \in J(v)} \frac{w(u)}{\beta(u)+1} \right\}$
6: Select $v^* = \max_{v \in Q} \frac{w(v)}{\beta(v)+1}$
7: Set $\mathbf{O} = \mathbf{O} \cup \{v^*\}$
8: Set $G = G - J(v^*)$
9: Output $\mathbf{O}$

---

here, $\mathbf{O}$ is the maximum weight independent set in the graph, which is the schedule pattern corresponding maximum weight sum data rate. $J(v)$ is the sub-graph of G containing vertex v and the vertices adjacent to v, $\beta(v)$ is the degree of v, which is the number of vertices adjacent to v. Q is the set of vertices where the weight of vertex v is larger than the average weight of $J(v)$. v^* is selected by making the average weight of $J(v)$ maximization.

9.5.3 Power Allocation

Once the user scheduling is determined, device power can be allocated according to the channel condition to achieve the maximum sum data rate. Power allocation in NOMA has been extensively investigated in the existing works. To achieve the maximum sum data rate under fairness constraints, a similar algorithm to [149] is used here. We notice that the objective function (9.8a) as a logarithmic function of SINR is monotonically increasing. It can be transformed into a product of exponential linear fraction functions. Due to the properties of logarithm function, the optimal power allocation problem for a specified user combination is

$$\max \prod_{k=1}^{K} (\frac{\mu_k(\mathbf{p})}{\phi_k(\mathbf{p})})^{w_k}, \tag{9.11a}$$

$$\text{s.t.} \quad 0 \le p_k \le p_k^{\max}, \forall k \in \mathcal{K}. \tag{9.11b}$$

where $\mathbf{p} = (p_k, \forall k \in \mathcal{K})$ is the power vector, $\mu_k(\mathbf{p}) = \sum_{j=k}^{K} p_j h_j^2 + \sigma^2$ and $\phi_k(\mathbf{p}) = \sum_{j=k+1}^{K} p_j h_j^2 + \sigma^2$. Let $\mathbf{z}_k = \frac{\mu_k(\mathbf{p})}{\phi_k(\mathbf{p})}$ for all k, the problem then can be re-formulated as

$$\max \prod_{k=1}^{K} (\mathbf{z}_k)^{w_k} \tag{9.12a}$$

$$\text{s.t.} \quad 0 \le \mathbf{z}_k \le \frac{\mu_k(\mathbf{p})}{\phi_k(\mathbf{p})}, \forall k \in \mathcal{K}, \tag{9.12b}$$

$$0 \le p_k \le p_k^{\max}, \forall k \in \mathcal{K}. \tag{9.12c}$$

Notice that $\tau(\mathbf{e}) = \prod_{k=1}^{K} (e_k)^{w_k}$ is an increasing function for all positive e_k, where $\mathbf{e}$ is the collection of all e_k. Besides, for two vectors $\mathbf{e}_l$ and $\mathbf{e}_m$, if $\mathbf{e}_l \succ \mathbf{e}_m$, where $\succ$ means element-wise greater than, we have $\tau(\mathbf{e}_l) > \tau(\mathbf{e}_m)$. Clearly, the optimal solution occurs where $\mathbf{z}_k^* = \frac{\mu_k(\mathbf{p}^*)}{\phi_k(\mathbf{p}^*)}$, and p_k in the feasible set. This can be regarded as a multiplicative linear fractional programming (MLFP) problem, where K linear equations are formulated as below:

$$\mathbf{z}_k^* \phi_k(\mathbf{p}^*) - \mu_k(\mathbf{p}^*) = 0, \forall k \in \mathcal{K}. \tag{9.13}$$

Notice that (9.13) contains random channel gain components; hence, those K linear equations are independent with probability 1, which suggests a unique optimal power allocation $\mathbf{p}^*$. To solve (9.13) efficiently, however, requires constructing of feasible polyblock and sequentially reduce its size, see [149] for the detailed algorithm.

9.6 Numerical Results

In this section, we present the experimental FL update results for both TDMA-based original FedAvg [128] and NOMA compression-based FedAvg schemes by using diverse learning models and federated datasets. The channel parameters are given as follows. The uplink bandwidth is $B = 5$ MHz, path loss exponent $\alpha = 3$, additive noise power density $\sigma^2 = -174$ dBm/Hz. The number of the selected users is set as two values, $K = 10$ and $K = 20$, for different runs. All the users are randomly distributed in a disk region with a radius 500m and they have the same transmission power $p_k = 0.1$ watts. Uplink transmission time slot is 0.5 s. Downlink transmission from the PS to all the users uses broadcast and is uncompressed. The transmission time is calculated as $T_d = \max_k \frac{32*P}{B_d \log_2(1+P_d \gamma_k)}$, where P is the number of the total updated parameters, B_d is the downlink bandwidth at 10 MHz. $P_d = 2$ watts is the PS power, γ_k is the SNR from the PS to k-th user. To prove the generality of the proposed scheme, we use convex loss function on image classification problems with MNIST (Modified National Institute of Standards and Technology database, a database of handwritten digit images) and Federated Extended MNIST (FEMNIST) datasets and non-convex loss function on text sentiment analysis task on tweets from Sentiment140 (*Sent140*). The convex problem is solved with LeNet-300-100 model, which is a fully connected network with two hidden layers. The first layer consists of 300 neurons and the second layer consists of 100 neurons. The non-convex problem uses a long short-term memory (LSTM) classifier. We further employ the datasets from [103] so that the data points on different devices are non-i.i.d., i.e. the number of data points and their high-level model representations vary across different devices. Each device trains the learning model individually and then uploads the model parameters to the PS for the weighted average. The statistics of the datasets are summarized in Table 9.2.

For hyperparameters in FL, we use batch size $B = 10$ on all datasets. The learning rate and number of communication rounds are fixed for each dataset but may vary among different datasets. Specifically, we use learning rate $\eta = 0.001$ and maximum communication round $T = 100$ for MNIST, $\eta = 0.003$, $T = 300$ for FEMNIST, and $\eta = 0.05$, $T = 100$ for *Sent140*. For each user, it uses the majority portion of the data for training and the rest for testing. Testing is performed at each device and accuracy is calculated by taking the average across all the devices.

Figure 9.4 presents the test accuracy result from the MNIST dataset. In the proposed scheme, NOMA is used in the uplink channel and the model parameters are compressed with either adaptive quantization or sparsification. The average compression ratio for adaptive quantization at each round is 0.55 for $K = 10$ and 0.33 for $K = 20$. For adaptive

Table 9.2 Statistics of datasets.

Dataset	No. of arameters (P)	No. of devices (N)	No. of data (D)
MNIST	266,610	1,000	69,035
FEMNIST	266,610	200	18,345
Sent140	243,861	660	40,783

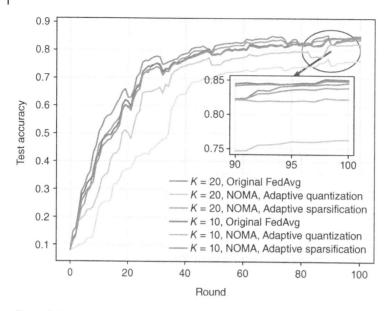

Figure 9.4 Test accuracy comparison under different scenarios, when $K = 10$, $K = 20$, and original TDMA-based FedAvg with our proposed NOMA compression-based FedAvg.

sparsification, the average compression ratio is 0.53 and 0.31. As a comparison, original FedAvg algorithm [128] with no gradient compression (32-bit per parameter) is also implemented with TDMA as the access scheme. It can be readily observed that all the schemes except the adaptive quantization with $K = 20$ users achieve a similar accuracy (over 80%) at round 100. The reason is that, when $K = 20$ users participate the model updates simultaneously with NOMA, mutual interference within each round causes significant data rate degradation for the first few decoded users, which then leads to a more aggressive compression strategy, especially for quantization where most of the parameters are set to 0.

While Figure 9.4 shows that the proposed scheme has a comparable accuracy with the original FedAvg. For simplicity, we only show the scenario with $K = 10$. With the simulation settings, each round for NOMA and compression-based FedAvg scheme corresponds to $t_k + T_d$ second and each round for the original FedAvg scheme corresponds to $Kt_d + T_d$ seconds. Thus NOMA and compression-based FedAvg scheme takes around 70s time to achieve 85% accuracy. In Figure 9.5, in order to achieve the same accuracy, the original TDMA-based FedAvg takes more than 500 s. NOMA-aided FL can save 7.4× communication time during the update process. Alternatively, with 500 s training in NOMA-based protocol, we see the accuracy improves only from 85% to 88.6% for adaptive quantization and to 90% for adaptive sparsification. Notice that under $K = 20$, the time difference is more prominent.

The proposed NOMA-enabled adaptive compression scheme is also proved to be effective for non-i.i.d. FEMNIST dataset (Figure 9.6) and non-convex loss functions for *Sent140* (Figure 9.7). Under all scenarios, we observe test accuracy fluctuations, especially for the FEMNIST dataset as it is a highly non-i.i.d. scenario and almost every device has a distinct data distribution. Nevertheless, one can observe similar performance between the proposed

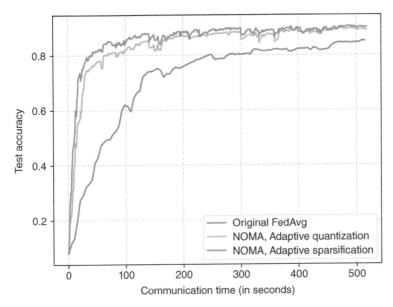

Figure 9.5 Test accuracy comparison between original TDMA-based FedAvg and NOMA compression-based FedAvg update with communication time.

Figure 9.6 Test accuracy on FEMNIST datasets: Test accuracy comparison vs communication rounds.

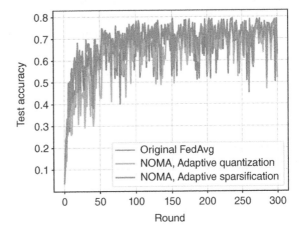

algorithm and the original FedAvg in terms of accuracy. Notice that the original FedAvg and our adaptive sparsification scheme have almost identical result at each round; hence, they can be hardly differentiated from the curves. Again the adaptive quantization gives a relatively worse performance than other two schemes.

From latency performance perspective, in the final round of FEMNIST dataset, to obtain 79.5% of accuracy, TDMA-based original FedAvg takes approximately 1600 s, while both NOMA-enabled adaptive compression schemes consume around 200 s. Similarly, *Sent140* achieves 73.5% accuracy with 75 s with our scheme but over 510 s with TDMA-based FedAvg. These results have again proved the remarkable low latency performance of the proposed algorithm.

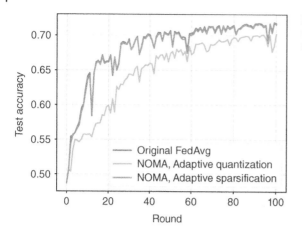

Figure 9.7 Test accuracy on *Sent140* datasets: Test accuracy comparison vs communication rounds.

9.7 Summary

In this work, we proposed to apply NOMA in the uplink FL model update. We considered wireless fading channels during the update process and adaptively compressed gradient values according to either sparsification or quantization method. To further improve system performance, we designed user scheduling and power allocation algorithm. Simulation results from three different datasets demonstrated that the proposed scheme can significantly reduce the communication latency without any compromise on the test accuracy.

10

Secure Spectrum Sharing with Machine Learning: An Overview

10.1 Background

The exponential growth of Internet-connected systems has generated numerous challenges for current and future wireless systems. One main issue is spectrum shortage. Even though Chapters 2–9 have discussed using 5G NR technology (mainly NOMA), as well as CR to alleviate this challenge, it is still urgent to call for an adaptive, reliable, and scalable spectrum sharing (SS) mechanism, such that the scarce spectrum resources can be efficiently utilized. Furthermore, the complicated communication environment brings more risks to the users and systems. Thus, security and privacy issues have become the primary concern in 5G networks. As an example, the recent spectrum battle between federal aviation administration (FAA) and several Internet service providers (ISPs) over 5G C-band has brought attention on the radio security for airplanes when 5G is in active deployments [195].

Different from traditional exclusive frequency allocations, SS by definition involves multiple entities and uses the spectrum in a shared way in order to increase the efficiency of the limited spectrum resources. According to [40], SS can fall into two main categories: horizontal sharing and vertical sharing. In horizontal sharing, it implies that all the networks and users have equal rights to access the spectrum. Such methods allow the users to co-exist peacefully and efficiently. Vertical sharing, on the other hand, allows multi-type users to access the spectrum resources with different rights. Therefore, secondary users (SUs) can use the spectrum without harming the performance of primary users (PUs). By enabling SUs to access the spectrum owned by the PUs, the limited spectrum resource can support more devices.

One of the technical challenges in the vertical SS system is how to guarantee the performance of different types of users while achieving the highest SE. To this end, the spectrum access mechanism, interference control, resource allocation, and fairness all need to be tackled in a dynamic and collaborative way. Since the concept of SS was first introduced, different SS frameworks based on various application scenarios have been developed by researchers.

This chapter provides a comprehensive and systematical review on secure SS in 5G and beyond systems. The focus is more at introducing the machine learning (ML)-based methodologies, such as deep reinforcement learning.

5G and Beyond Wireless Communication Networks, First Edition. Haijian Sun, Rose Qingyang Hu, and Yi Qian.
© 2024 John Wiley & Sons Ltd. Published 2024 by John Wiley & Sons Ltd.

10.1.1 SS: A Brief History

The concept of vertical SS was first brought up by decentralized and opportunistic cognitive radio (CR) techniques, where SUs exploited the idle spectrum with sensing ability to transmit their information without causing any harmful interference to the licensed PUs [132].

Traditional CR techniques enabled SUs to take advantage of spectrum opportunities by learning/monitoring the environment and adjusting their transmission parameters adaptively. However, the opportunistic access to the bands without a license in CRN makes it challenging to guarantee the QoS and maintain a low level of interference when multiple service providers co-exist. In particular, with the increase of primary wireless devices and activities in the network, SUs may have very limited access opportunities to the spectrum resource. Therefore, new licensed spectrum access methods are needed to provide more predictable and controllable solutions to meet the high QoS requirement for both SUs and PUs [183].

In 2012, Qualcomm and Nokia initially proposed the concept of Authorized Shared Access (ASA), which was further extended to the Licensed Shared Access (LSA) framework by several European institutions, including the European Conference of Postal and Telecommunications Administrations (CEPT), the European Commission (EC), and the Radio Spectrum Policy Group (RSPG) of the European Union (EU). The main objective of LSA is to allow new users to work in already-occupied frequency bands while maintaining existing incumbent services on a long-term basis. The LSA framework is currently switched to explore the 3.4–3.8 GHz band from the 2.3–2.4 GHz band, enabling coexistence between incumbents and 5G applications [127, 144].

Similar to the LSA framework, the US Federal Communications Commission (FCC) proposed the Citizens Broadband Radio Service (CBRS) in order to open and share the frequency band 3.55–3.7 GHz. It sought to improve spectrum usage by allowing commercial users to share the band with incumbent military radars and satellite earth stations. In this approach, the access and user coordination are controlled by the corresponding Spectrum Access System (SAS). SAS comprises three types of users with different levels of priority. The first type is Incumbent Users (IUs), which have the highest spectrum access priority. The second is the Priority Access License (PAL) users, which can exclusively access the spectrum without the existence of IUs. The third is General Authorized Access (GAA) users, who have sensing-assisted unlicensed access in the absence of the incumbent and PAL users. SAS determines the maximum allowable transmission power level and the available frequencies at a given location to be assigned to PAL and GAA users [127, 144].

It should be noted that both LSA and SAS systems are defined for usage in a specific frequency band. LSA is mainly based on database-assisted SS, while SAS combines a database with Environmental Sensing Capability (ESC) protections. Thus in SAS, the radio resources allocation decision is obtained with assistance from the spectrum database and sensing results. The ESC can more effectively protect IUs from harmful interference while guaranteeing their privacy. The database can provide a more stable service for SUs than CRN [75].

Much research has been devoted to increasing licensed systems' capacities by the extension of Long-Term Evolution (LTE) over the unlicensed spectrum band. Several concepts have been proposed, such as LTE in unlicensed bands (LTE-U), License Assisted Access (LAA) in LTE Advanced (LTE-A), and LTE Wireless Local Area Network (WLAN)

Aggregation (LWA). By allowing LTE users to operate on the unlicensed band without causing any harmful interference to original users such as WiFi devices, coexisting technologies can have a great impact on the spectrum access in the immediate future [144].

The emergence of IoT networks presents new challenges to wireless communication design from both spectrum and energy aspects. Supporting communication with power-limited IoT devices, Ambient Backscatter Communication (AmBC) has attracted extensive attention as a promising technology for SS communications. In AmBC systems, backscatter devices can use surrounding signals from ambient RF sources to communicate with each other. By modulating and reflecting surrounding ambient signals, the backscatter transmitter can transmit data to the receiver without consuming new spectrum resources. The receiver can decode and obtain useful information after receiving the signal. Therefore, the AmBC system does not require a dedicated frequency spectrum, and the number of RF components is minimized at backscatter devices. Those devices can transmit data with sufficient harvested energy from RF sources [188], which can also improve system energy efficiency significantly.

In summary, as shown in Figure 10.1, spectrum sharing originated from the concept of opportunistic access. Database-supported access frameworks on a specific licensed frequency band were then developed to connect new users to the unused licensed band without degrading the performance of IUs to improve the SE. As the number of devices as well as the demand for network capacity has increased, the expansion of licensed band services to the unlicensed band has been proposed. LTE-U provides a view of how to use the unlicensed band to improve the licensed users' performance. Finally, symbiotic schemes such as AmBC help researchers deal with the massive growth of IoT devices that normally have power and resource limitations, providing a new paradigm for spectrum sharing. Although these research studies share some overlapping features such as sensing and access control, they each have their distinctive focuses. 5G has a very broad technical scope and needs to address a variety of device communication problems. Therefore, investigating SS issues under different frameworks can provide very instrumental insights for future communication development.

With the advancement of computing technologies such as GPU and algorithm development, ML has garnered tremendous interest and recently demonstrated astonishing potential for tackling large-scale, highly dynamic, very complicated problems that traditional techniques cannot easily handle. ML algorithms have gained advantages in processing, classification, decision-making, recognition, and other problems [17].

SS frameworks naturally share common features with ML, making the combination of ML with SS networks very appealing. For all the frameworks mentioned above, users/ coordinators in the SS network need to observe the spectrum resource usage and make corresponding decisions in accordance with the three main conditions for intelligence,

Figure 10.1 Spectrum sharing paradigm.

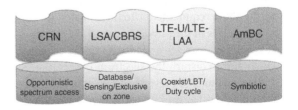

i.e. perception, learning, and reasoning [17]. In ML, the intelligent agent first senses the surrounding environment and internal states through perception to obtain information. It further transforms that information into knowledge by using different classification methodologies and generalizes the hypothesis. Based on the obtained knowledge, it then achieves certain goals through reasoning.

The combination of SS with ML techniques has the potential to adaptively tackle complicated and dynamic allocation and classification problems such as channel selection, interference control, and resource allocation. Applying ML to different SS frameworks has become a research hotspot and a promising frontier in the field of future communications. Toward that end, this study aims to provide a timely and comprehensive survey in this up-and-coming field.

10.1.2   Security Issues in SS

The development of SS techniques will help relieve spectrum scarcity. However, due to the dynamic access of spectrum resources by a variety of users, SS systems can be exposed to malicious attackers. In the first place, the lack of ownership of the spectrum leaves unlicensed users highly susceptible to malicious attacks. Therefore, it is hard to protect their opportunistic spectrum access from adversaries. In the second, the dynamic spectrum availability and distributed network structures make it challenging to implement adequate security countermeasures. Moreover, in some SS systems, PUs may contain sensitive information, which can be effortlessly obtained by malicious SUs during the SS process. Finally, new technologies such as ML may also be exploited by attackers with even more complicated and unpredictable attacks [206]. Clearly, security issues are of great concern and impose unique challenges in the SS network.

According to [145], security requirements in most of the spectrum-sharing scenarios include confidentiality, integrity, availability, authentication, non-repudiation, compliance, access control, and privacy. Confidentiality means sensitive information should not be disclosed to unauthorized users, especially in database-assisted SS systems. Integrity ensures that information communicated among users is protected from malicious alteration, insertion, deletion, or replay. Availability assures users access to the spectrum/database when it is required. Authentication requires that the users should be able to establish and verify their identity. Non-repudiation means users should be able to deny having received/sent a message or to deny having accessed the spectrum at a specified location and time. Compliance means the network should be able to detect non-compliant behavior that results in harmful interference. Access control indicates that users should not access the spectrum/database without credentials. Privacy means users' sensitive/private information should be protected.

Diverse security threats in different network layers can prevent the SS system from meeting the above requirements. In this chapter, we will mainly focus on the threats and mitigation strategies in the physical layer of the SS network. We investigate works related to two classical spectrum sensing attacks in the SS network, i.e. PUE attacks and SSDF attacks, which aim to disturb the spectrum observation and users' access to the system. We also studied methods of preventing two attacks that commonly exist in wireless communication networks, i.e. jamming attacks and eavesdropping attacks. The special features of the

SS network provide new defense solutions for these common attacks. Since PUs need to open their exclusive license spectrum to coexist with multi-type users in some SS frameworks, we also investigated privacy issues and corresponding countermeasures.

To further enhance the security performance of the SS system, ML has become an important part of security and privacy protections in various applications. ML is a powerful tool for data exploration and can distinguish normal and abnormal behaviors based on how devices in the SS system interact with each other during spectrum access. The behavioral data of each component in the SS network can be collected and analyzed to determine normal patterns of interaction, thereby allowing the system to identify malicious behaviors early on. Furthermore, ML can also be used to intelligently predict new attacks, which often are the mutations of previous attacks by exploring the existing records. Consequently, the SS network must transition from merely facilitating secure communication to security-based intelligence enabled by ML for effective and secure systems. The state-of-the-art learning-based security solutions for SS systems will also be comprehensively reviewed in this chapter.

10.2 ML-Based Methodologies for SS

In this section, key processes in the CRN network are first introduced, which are shared by most spectrum frameworks. Summarizing the roles that ML plays in these processes can help us to better understand the combination of ML and other frameworks. It also allows us to better assess the security risks in existing SS frameworks with a comprehensive understanding of the mechanism behind each SS technique. The special problems faced by database-assisted SS frameworks (such as LSA and CBRS) are then investigated, i.e. how to protect IUs when unlicensed users are introduced into licensed spectrum bands. Next, a discussion of the application of ML to the coexistence of licensed LTE systems and unlicensed WiFi systems in unlicensed frequency bands is presented. Finally, a comprehensive study of the AmBC system using ML is conducted to gain insight into the symbiosis-based SS framework as well as the benefits of the combination of AmBC and CRN. The content structure is as illustrated in Figure 10.2.

10.2.1 ML-Based CRN

In CRN, unlicensed SUs need to identify the vacant or unoccupied licensed frequency band (spectrum hole) owned by licensed PUs [73]. After spotting the spectrum hole, SUs can access it without visibly interfering with any PU. If a PU's activity reappears, the SUs must vacate the spectrum immediately. This dynamic and uncertain environment creates unique and complex challenges within the CRN. However, ML algorithms are very effective in dealing with such challenges and can help improve system performance.

As shown in Figure 10.3, the major steps in CRN can be summarized as spectrum sensing, spectrum selection, spectrum access, and spectrum handoff [2]. The CR agent first uses the sensing function to monitor the unused spectrum and search for possible access opportunities for SUs. Based on the sensing results, the spectrum selection function helps SUs select the best available channels and the spectrum access mechanisms provide fair spectrum

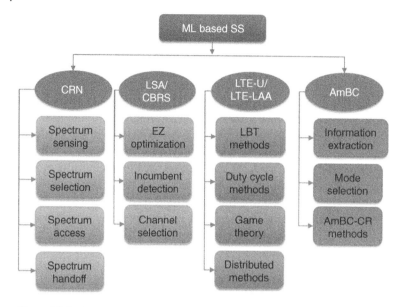

Figure 10.2 ML-based methodologies for SS.

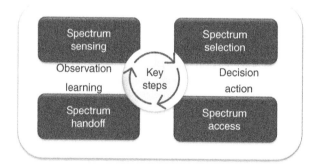

Figure 10.3 Key steps in CRN.

scheduling among vying SUs. Since a channel must be vacated when the PU reappears, the corresponding SU must perform a spectrum handoff function to switch to another available channel or wait until the channel becomes idle again. It is worth noting that most of the existing vertical SS approaches adopted these four steps in their frameworks.

10.2.1.1 Spectrum Sensing

Before an SU accesses the licensed channel, it needs to first observe and measure the state of the spectral occupancy (i.e. idle/busy) by performing spectrum sensing. During this procedure, the SU needs to distinguish the signal of PUs from background noise and interference. As such, spectrum sensing can be formed as a classification problem.

Automatic modulation recognition is a keystone of CR adaptive modulation and demodulation capabilities to sense and learn environments and make corresponding adjustments. Automatic modulation recognition can be deemed equivalent to a classification problem, and deep learning (DL) achieves outstanding performance in various classification tasks. Several existing works investigate the combination of DL with automatic modulation recognition in CRNs.

In [130], a DL-based automated modulation classification method that employed Spectral Correlation Function (SCF) was proposed. Deep Belief Network (DBN) was applied for pattern recognition and classification. By using noise-resilient SCF signatures and DBN that are effective in learning complex patterns, the proposed method can achieve high accuracy in modulation detection and classification even in the presence of environmental noises. The efficiency of the proposed method was verified in classifying 4FSK, 16QAM, BPSK, QPSK, and OFDM modulation based on various environments settings.

ML provides effective tools for automating CR functionalities by reliably extracting and learning intrinsic spectrum dynamics. However, there are two critical challenges. First, ML requires a significant amount of training data to capture complex channel and emitter characteristics and train the algorithm of classifiers. Second, the training data that has been identified for one spectrum environment cannot be used for another, especially when channel and emitter conditions change [39]. To address these challenges, various robust spectrum sensing mechanisms have been developed. A new approach to training data augmentation and domain adaptation was presented in [39]. A Generative Adversarial Network (GAN) with DL structures was employed to generate additional synthetic training data to improve classifier accuracy and adapt training data to spectrum dynamics. This approach can be used to perform spectrum sensing when only limited training data is available and no knowledge of spectrum statistics is assumed. Another robust spectrum sensing framework based on DL was proposed in [148]. The received signals at the SU's receiver were filtered, sampled, and then directly fed into a convolutional neural network (CNN). To improve the adaptive ability of the classifier, Transfer Learning (TL) was incorporated into the framework to improve robustness.

Besides improving the accuracy and robustness of spectrum sensing, another substantial sensing performance improvement comes from using ML to help the SU make efficient decisions regarding which channel to sense and when or how often to sense.

The prediction ability can enable SUs to perform spectrum sensing in a more efficient manner. By enabling SUs to determine the channel selection for data transmission and predicting the period of channel idle status, sensing time can be significantly reduced. Therefore, the authors in [152] proposed an ML-based method that employed a Reinforcement Learning (RL) algorithm for channel selection and a Bayesian algorithm to determine the length of time for which sensing operation can be skipped. It was shown that the proposed method could effectively reduce the sensing operations while keeping interference with PUs at an acceptable level. This work also showed that by skipping unnecessary sensing, SUs can save more energy and achieve higher throughput by spending the saved sensing time for transmission. A Hidden Markov Model (HMM)-based Cooperative Spectrum Sensing (CSS) method was proposed in [91] to predict the status of the network environment. First, the concept of an Interference Zone (IZ) was introduced to indicate the presence of PUs. Then, by combining the sensing results from SUs located in different IZs, the Fusion Center employed a fusion rule for modeling specific HMM. Moreover, the system adopted a Baum–Welch (BW) algorithm to estimate the parameters of the HMM-based past spectrum sensing results. The estimated parameters were then passed to a forward algorithm to predict the activity of PUs. Finally, SUs were classified into two categories according to the prediction results, i.e. Interfered by PU (IP) and Not Interfered by PU (NIP). SUs marked as IP do not need to perform spectrum sensing to avoid unnecessary energy consumption.

10.2.1.2 Spectrum Selection

After the system receives spectrum sensing results, the spectrum selection is performed to capture the best available spectrum to meet user needs. As a decision-making problem, it requires the system to adaptively capture the optimal choice based on observations of the environment. RL algorithms are appealing tools for designing systems that need to perform adaptive decision-making. In RL, a learner takes actions by trial and error, and learns the action patterns suitable for various situations based on the rewards obtained from these actions. Exploration actions are selected in situations even when the knowledge about the environment is uncertain. This mechanism fits into the spectrum selection problems.

As shown in Figure 10.4, at the beginning of the RL cycle, the agent receives a full or partial observation of current states and the corresponding reward. Combining those states and rewards, the policy is updated by each agent during the learning stage. Then the agent performs a certain selection action based on the updated policy at the decision stage. With RL, CRN can be modeled as a distributed self-organized multi-agent system in which each SU or agent performs spectrum selection by efficiently interacting with the environment through a learning policy. In this approach, other SUs' decisions can be considered as a part of the responses of the environment for each SU.

A distributed Learning Automata (LA)-based spectrum selection scheme was studied in [53]. It aimed to enable the SUs to sense the RF environment intelligently and to learn its different responses. On the other hand, the activity information of PUs and other SUs was not available, and different SUs could not exchange their information. The self-organized SUs performed the channel selection as the action and received the corresponding response indicating how favorable the action was. Based on the response, SUs could determine the optimal spectrum selection to achieve a lower transmission delay and lower interference to PUs and other SUs.

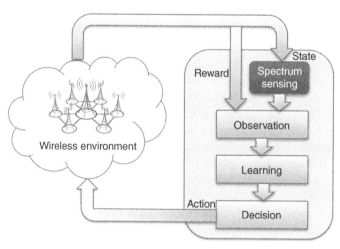

Figure 10.4 The reinforcement learning cycle.

Two Q-learning-based spectrum assignment methods were developed in [239]. The independent Q-learning-based scheme was designed for a case in which the SUs could not exchange information, while the collaborative Q-learning-based scheme was designed for a case in which information could be exchanged among SUs. It was shown that the collaborative Q-learning-based assignment method performed better than the independent Q-learning-based method.

Most cases fell short of the perfect CSI, and imperfect CSI can degrade the performance of system spectrum selection. To overcome this challenge, an RL-based robust decentralized multi-agent resource allocation scheme was proposed in [95]. They introduced cloud computing to enlarge storage space, reduce operating expenditures, and enhance the flexibility of cooperation. A cooperative framework in a multi-agent system was then developed to improve the performance of their proposed scheme in terms of network capacity, outage probability, and convergence speed.

To better manage users in a CRN, the clustering operation organizes SUs into logical groups based on their common features. Clustering can provide network scalability, spectrum stability, and fulfill cooperative tasks. In each cluster, one SU can either act as a leader (or cluster head) that manages essential CR operations, such as channel sensing and routing, or as a member node that associates itself with the cluster head. Cluster size represents the number of nodes in a cluster and affects various performance metrics. In the CRN, the cluster size adjustment and cluster head selection can significantly impact system performance.

In [89], the authors proposed a first-of-its-kind cluster size adjustment scheme based on RL. The proposed scheme adapts the cluster size according to the number of white spaces to improve network scalability and cluster stability. It was shown that their proposed scheme improved network scalability by creating larger clusters and improved cluster stability by reducing the number of re-clusterings (the number of cluster splits) and clustering overhead while reducing interference between licensed and unlicensed users in CRNs.

Using Q-value to evaluate the channel quality in Cluster-based CR Ad-Hoc Networks (CRAHN), a Q-learning-based cluster formation mechanism was studied in [74]. Channel quality, residual energy, and network conditions were jointly considered to form a distributed cluster network. All the nodes built their neighboring topology by exchanging the channel status and neighbor list information. Each node then selected the optimal cluster head candidate. Distributed cluster head selections, optimum common active data channel decisions, and gateway node selection procedures were presented. It was shown that the proposed approach could extend the network lifetime and enhance reachability.

10.2.1.3 Spectrum Access

One important question in CRNs spectrum access is how to assign limited resources, such as available spectrum channels and transmit powers, to maximize the system throughput and efficiency. Numerous related works based on RL [94, 147, 241], DL [113, 115], and Deep RL (DRL) [101, 161] have been carried out.

An RL-based resource allocation approach entitled Q-Learning and State-Action-Reward-State-Action (SARSA) was proposed in [94] that mitigated interference without the requirements of the network model information. Users in this method act as multiple agents and cooperate in a decentralized manner. A stochastic dynamic algorithm was formed to determine the best resource allocation strategy. It was shown that the energy efficiency could be significantly improved by the proposed approach without sacrificing user QoS.

An energy harvesting-enabled CRN was investigated in [147]. To achieve higher throughput, a harvest-or-transmit policy for SUs transmit power optimization was proposed. A Q-learning-based online policy was developed first to deal with the underlying Markov process without any prior knowledge. An infinite horizon stochastic dynamic programming-based optimal online policy was then proposed by assuming that the full statistical knowledge of the governing Markov process was known. Finally, a generalized Benders decomposition algorithm-based offline policy was given, where the energy arrivals and channel states information were known before all transmitters for a given time deadline.

By combining Multi-Armed Bandit (MAB) and matching theory, the ML-assisted Opportunistic Spectrum Access (OSA) approach was developed in [241]. A single SU case was first considered without the volatility of channel availability information. Next, the upper confidence bound algorithm-based Occurrence-Aware OSA (OA-OSA) framework was designed to achieve the long-term optimal network throughput performance and the trade-off between exploration and exploitation. The OA-OSA was then extended to the multi-SU scenario with channel access competitions by integrating the Gale–Shapley algorithm.

In [115], by considering the OFDMA-based resource allocation for the underlying SUs, researchers aimed to minimize the weighted sum of the secondary interference power under the constraints of QoS, power consumption, and data rate. A Damped Three-Dimensional (D3D) Message-Passing Algorithm (MPA) based on DL was proposed, and an analogous back-propagation algorithm was developed to learn the optimal parameters. A sub-optimal resource allocation method was developed based on a damped two-dimensional MPA to improve computational efficiency. By considering the EE and SE, as well as Computing Efficiency (CE) for both PUs and SUs, a DL-based resource allocation algorithm in CRNs was proposed in [113] to minimize the weighted sum of the secondary interference power. It was shown that the proposed scheme significantly improved both the SE and EE for PUs and SUs.

Insufficient specificity and function approximation can impose some limitations on RL algorithms, but neural networks can be used to compensate for them. DRL algorithms are capable of combining the process of RL with deep neural networks to approximate the Q action-value function. Compared with conventional RL, DRL can significantly improve learning performance and learning speed. DRL has attracted a lot of attention in research for solving the problems in CR networks such as resource allocation, spectrum management, and power control.

In [161], the authors presented a DRL-based resource allocation method for CRN to maximize the secondary network performance while meeting the primary link interference constraint. By adopting a Mean Opinion Score (MOS) as the performance metric, the proposed model seamlessly integrates resource allocations among heterogeneous traffic. The resource allocation problem was solved by utilizing a Deep Q Network (DQN) algorithm where a neural network approximated the Q action-value function. TL was incorporated into the learning procedure to further improve the learning performance. It was shown that TL reduced the number of iterations for convergence by approximately 25% and 72% compared to the DQN algorithm without utilizing TL or standard Q-learning, respectively.

10.2.1.4 Spectrum Handoff

Spectrum handoff is intended to maintain seamless communication during the transition to a better spectrum. However, enabling spectrum handoff for multimedia applications in a CRN is challenging due to multiple interruptions from PUs, contentions among SUs, and heterogeneous Quality-of-Experience (QoE) requirements. Although an SU may not know exactly when the PU comes back, it always wants to achieve reliable spectrum usage to support the QoS requirements. If the quality of the current channel degrades, the SU can make one of the following three decisions:

(1) Stay in the same channel and wait for it to become idle again (called stay-and-wait).
(2) Stay in the same channel and adapt to the varying channel conditions (called stay-and-adjust).
(3) Switch to another channel that meets the QoS requirement (called spectrum handoff).

In [207], a learning-based and QoE-driven spectrum handoff scheme was proposed to maximize the multimedia users' satisfaction. A mixed preemptive and non-preemptive resume priority (PRP/NPRP) M/G/1 queueing model was designed for the spectrum usage behaviors of prioritized multimedia applications. The RL-assisted QoE-driven spectrum handoff scheme was developed to maximize the quality of video transmissions in the long term. Their proposed learning scheme could adaptively perform spectrum handoff based on the variation of channel conditions and traffic loads.

To address limitations of PRP/NPRP queuing models, the authors in [210] employed a hybrid queuing model with discretion rules to characterize the SUs' spectrum access priorities. The channel waiting time during spectrum handoff was then calculated according to this hybrid queuing model. The multi-teacher knowledge transfer method was further proposed to accelerate the algorithm, wherein the multiple SUs that already had mature spectrum adaptation strategies could share their knowledge with an inexperienced SU.

10.2.2 Database-Assisted SS

The sensing-driven OSA system aims to explore the spectrum holes in the unlicensed band, which can be inefficient and unreliable, especially when the number of wireless communication devices increases rapidly. To better serve the secondary system, a specific spectrum band like 3.5 GHz is opened to the public. Database-assisted Dynamic SS (DSS) systems such as LSA and SAS have been proposed to better coordinate the users in different systems with various spectrum access priorities.

LSA has two types of users: incumbents and SUs, where incumbents send their spectrum usage information to a database center called LSA Repository [136]. The system then decides whether the SUs can access the spectrum resource with this information or not and no sensing ability is required for those users. The SAS system also maintains a similar database for three different types of users. The difference is that the IUs in the SAS system may have very sensitive information and do not want to offer it to the database [136]. Instead, to protect the IUs, the Exclusion/Protection Zone (EZ) is applied where the SUs (PAL and GAA users) are banned from accessing the spectrum in these areas to prevent them from harmful interference to IUs. Environment sensing capability (ESC)-based incumbent detection also requires users in tiers 2 and 3 to perform sensing. The system then decides the spectrum access based on the EZ and ESC nodes' sensing results [14].

The existing ML-based works for database-assisted SS networks are mainly focused on the EZ adjustment [71, 218], ESC performance improvement [185, 202], and spectrum access coordination [29, 182, 226, 228].

10.2.2.1 ML-Based EZ Optimization

In a database-driven SS network, even with the SS policy provided by the database, harmful interference still can occur between a PU and an SU due to unexpected propagation paths. In a common solution, a primary EZ is provided for the PUs that prevents SUs from using the same spectrum in EZ to keep interference at an acceptable level. However, the size of EZ needs to be optimized to efficiently cover the regions where interference may occur and a region's shape is usually not circular.

In [218], an ML-based framework was developed to deal with the interference by dynamically adjusting the EZ. By considering the propagation characteristics and shadow fading, the framework employed under-sampling and over-sampling schemes to solve an imbalanced data problem which can degrade the estimation accuracy of the appropriate shape of EZ. It was shown that their proposed method could significantly reduce the area of the EZ by 54% compared with the fixed circular EZ setting. Furthermore, the proposed sampling scheme could achieve a 1% interference probability with 21% fewer iterations and a 6% smaller area compared with the existing sampling benchmarks.

Using VHF-band radio sensors and the ML technique, an outdoor location estimation scheme of a high-priority DSS system was proposed in [71]. The delay profiles measured in the very high frequency (VHF) band were employed to estimate location. The precision of the EZ could then be improved based on the estimated location of the PUs. By using the ARIB STD-T103 system operating in the VHF-band, they measured delay profiles in a mountainous environment in Japan with the Deep Neural Network (DNN). With the trained DNN, the location cluster of the high-priority terminal could be predicted without GPS by simply measuring the delay profile of the PUs. It was shown that their method could significantly improve the total correct localization rate by up to 80.0%.

10.2.2.2 Incumbent Detection

According to the FCC, the IUs in the CBRS band include authorized federal users such as U.S. Navy shipborne SPN-43 air traffic control radar operating in the 3550–3700 MHz band, Fixed Satellite Service (space-to-Earth) earth stations operating in the 3600–3650 MHz band, and for a finite period, grandfathered wireless broadband licensees operating in the 3650–3700 MHz band. Due to security restrictions, the SAS cannot access the information of those IUs. To alleviate harmful interference from PAL and GAA users, the Environmental Sensing Capability (ESC) enabled by sensor networks could detect transmissions from the Department of Defense radar systems and transmitted that information to the SAS. The SAS could then assign the spectrum resources to users with different priorities dynamically. However, the single sensor detection lacked precision due to its geolocation, while distributed multiple sensor networks led to a high information exchange overhead. Moreover, the extreme operational characteristics of incumbent military wireless applications could overwhelm the existing spectrum sensing methods. Several studies have sought to address these issues.

An ML-based spectrum sensing system called Federated Incumbent Detection in CBRS (FaIR) was proposed in [185]. FL was adopted for ESCs to collaborate and train a data-driven ML model for IU detection with minimal communication overhead. Unlike a naive distributed sensing and centralized model framework, their proposed method could exploit the spatial diversity of the ESCs and improve detection performance.

Because the existing ESC-based methods had the potential to incur a high communication overhead and lead to leakage of sensitive information, a compressed sensing (CS)-based FL framework was proposed in [202] for IU detection. To protect privacy, local learning models transmitted updating parameters instead of raw spectrum data to the central server. A Multiple Measurement Vector (MMV) CS model was further adopted to aggregate these parameters. Based on the aggregated parameters, the central server could gain a global learning model and send the global parameters back to local learning models. Their proposed framework could significantly improve communication and training efficiency while guaranteeing detection performance compared with the raw training sample method.

10.2.2.3 Channel Selection and Transaction

In the database-assisted SS system, each user needs to choose a proper vacant channel in order to avoid severe interference with others. When economic approaches are adopted to model the SS system, idle channels can be traded as commodities. For this reason, channel selection and spectrum trade problems are critical to the database-assisted SS system, and many works based on game theory have been proposed to solve them.

In [29], a database-assisted distributed white-space Access Point (AP) network design was studied. The cooperative channel selection problem was first considered to maximize system throughput, where all APs were owned by one network operator. A distributed channel selection problem was then formed between APs that belonged to different operators, and a non-cooperative state-based game was formulated by considering the mobility of SUs. It was also shown that this algorithm was robust to perturbation from SUs' leaving and entering the system.

In [226], a method of idle channels sharing in overlapped licensed areas among PUs and SUs was proposed. Based on supply and demand fluctuations in different areas, SUs were grouped based on their suppliers, and channel transaction quotas were set by PUs for these SU groups accordingly. By applying evolutionary games, the PUs could obtain quotas of Evolutionary Stable Strategy (ESS) to maximize their incomes. Furthermore, a learning process was designed for the PUs to attain the optimal realizable integer quotas.

10.2.3 ML-Based LTE-U/LTE-LAA

LTE-U has emerged as an effective technique for alleviating spectrum scarcity. Using LTE-U along with advanced techniques such as carrier aggregation can boost the performance of existing cellular networks. However, LTE was initially designed to operate in the licensed spectrum exclusively and was not for harmonious coexistence with other possible co-located technologies [180]. For this reason, introducing LTE into the unlicensed spectrum can cause significant coexistence issues with other well-established unlicensed technologies such as Wi-Fi, IEEE 802.15.4, or Bluetooth. To enable fair spectrum sharing with other technologies operating in the unlicensed spectrum, in particular with Wi-Fi,

new coexistence technologies are needed. On the other hand, not much research attention has been given to studying cooperation between the technologies. Networks that
participate in a cooperation scheme can exchange information directly or indirectly (via a
third-party entity) to improve the efficiency of spectrum usage in a fairway.

To standardize LAA technology in the 5 GHz spectrum, the 3GPP standardization group
aims to develop a single global framework of LTE in the unlicensed bands. The framework
should guarantee that the operation of LTE does not critically affect the performance
of WiFi networks. Initially they only considered the downlink operation LTE-A (LTE
Advanced) Carrier Aggregation (CA) in the unlicensed band. This was later expanded to
include the simultaneously operate downlink and uplink [146]. The LTE LAA employed a
Listen Before Talk (LBT) mechanism to avoid collision and interference between users.

LTE-U is another option for operating LTE in an unlicensed spectrum, where LTE base
stations exploit transmission gaps to facilitate coexistence with WiFi networks. The development of LTE-U technology is led by the LTE-U Forum, an industry alliance. LTE-U has
been designed to operate as an unlicensed LTE in countries where the LBT technique is
not mandatory, such as the United States and China. LTE-U defines the operation of primary cells in a licensed band with one or two secondary cells (SCells), every 20 MHz in
the 5 GHz unlicensed band: U-NII-1 and/or U-NII-3 bands, spanning 5150-5250 MHz and
5725-5825 MHz, respectively [146].

10.2.3.1 ML-Based LBT Methods

According to LTE LAA standards in 3GPP Release 13, the LTE system must perform the LBT
procedure (also known as Clear Channel Assessment, CCA) and sense the channel prior to
a transmission in the unlicensed spectrum. As shown in Figure 10.5, when the channel is
sensed to be busy, the LTE system must defer its transmission by performing an exponential
backoff. If the channel is sensed to be idle, it performs a transmission burst with a duration
from 2 to 10 ms, depending on the channel access priority class [124].

To exploit the benefits of communications in an unlicensed spectrum using LTE-LAA,
a DL approach for the resource allocation of LTE-LAA small base stations (SBSs) was
proposed in [28]. The proposed method employs a proactive coexistence mechanism that
enables future delay-tolerant LTE-LAA data requests to be served within a given prediction

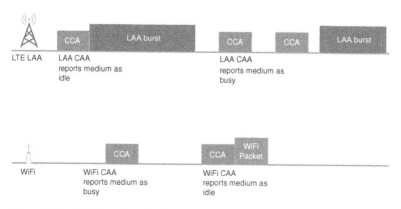

Figure 10.5 LBT-based method.

window before their actual arrival time. Therefore, it can improve the utilization of the unlicensed spectrum during off-peak hours while maximizing the total served LTE-LAA workloads.

To achieve long-term equal-weighted fairness between wireless local area networks and LTE-LAA operators, a non-cooperative game model was formulated where SBSs were modeled as homo egualis agents to predict a future action sequence. The proposed method enables multiple SBSs to proactively perform dynamic channel selection, carrier aggregation, and fractional spectrum access while guaranteeing equal opportunities for existing WiFi networks and other LTE-LAA operators.

10.2.3.2 ML-Based Duty Cycle Methods

Carrier Sensing Adaptive Transmission (CSAT) is a technique that can enable coexistence between LTE and Wi-Fi based on minor modifications of the 3GPP LTE Release 10/11/12 Carrier Aggregation protocols. As shown in Figure 10.6, CSAT introduces the use of duty cycle periods and divides the time into LTE "ON" and LTE "OFF" slots. During the LTE "OFF" period, also known as the "mute" period, LTE remains silent, gives other coexisting networks, such as Wi-Fi, an opportunity to transmit. During the LTE "ON" period, LTE accesses the channel without sensing it before transmission. Moreover, CSAT allows short transmission gaps during the LTE "ON" period to allow for latency-sensitive applications, such as VoIP in co-located networks. In CSAT, eNB senses the medium during a time period ranging from 10 to 100 ms and according to the observed channel utilization (based on the estimated number of Wi-Fi APs) defines the duration of the LTE "ON" and LTE "OFF" periods [180].

The existing work of LTE-U mainly focuses on using different RL algorithms to adjust the duty cycle and other network resources to maintain fairness between LTE and WiFi users, as well as to seek for a higher system capacity performance.

To investigate the application of LTE-U technology in the 3.5 GHz CBRS band, an MAB-based SS technique was developed in [146] for a seamless coexistence with WiFi. Assuming LTE-U to operate as a GAA user, they used MAB to adaptively optimize the duty cycle of LTE-U transmissions. Downlink power control was incorporated to achieve high EE and interference suppression. The study showed significant improvement in the aggregate capacity and cell-edge throughput of coexisting LTE-U and WiFi networks for different base station and user densities.

10.2.3.3 Game-Theory-Based Methods

The coexistence of LTE systems and WiFi system can usually be formed as a game theory problem, where each part must compete for the same unlicensed spectrum resource and finally reach an agreement with each other. Some game-theory-based LTE-U works are discussed in this section.

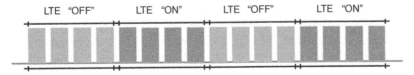

Figure 10.6 Duty cycle-based method.

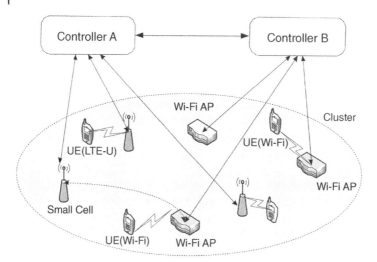

Figure 10.7 The coexistence of LTE-U and WiFi in an unlicensed spectrum. Source: [26] (© [2016] IEEE).

An SS scheme adapted for the operation of LTE-U and WiFi systems was proposed in [26]. The decision tree learning and repeated game were adopted to optimize unlicensed spectrum resource. As shown in Figure 10.7, the control plane was decoupled from the data plane and provided the system with a great capacity for processing data. Controllers learned the latest dataset in the pool to build decision trees and deduce the network status of the opponent. Repeated games for sharing spectrum resources were then employed to maximize resource utilization in the coexistence system. An incentive mechanism was further included to increase operators' motivation to share their spectrum resources.

10.2.3.4 Distributed-Algorithm-Based Methods

In [31], the authors investigated the uplink–downlink decoupling resource allocation problem for LTE-U-enabled Small Cell Networks (SCNs). By incorporating user association, spectrum allocation, and load balancing, the problem was formulated as a non-cooperative game. An Echo State Network (ESN)-based distributed algorithm was developed to address this problem. It was shown that even with limited information on the network's and users' states, the proposed algorithm could help the SBS to choose their optimal resource allocation strategies autonomously. Furthermore, the proposed method could significantly improve the sum rate of the 50th percentile of users and achieve a 167% increase compared to a Q-learning algorithm.

In [32, 33], a cache-enabled Unmanned Aerial Vehicles (UAVs) communication network serving wireless ground users over the LTE-U bands was considered. The problem under investigation was joint caching and resource allocation. By jointly incorporating user association, spectrum allocation, and content caching, a resource allocation problem was formed and a distributed algorithm based on the Liquid State Machine (LSM) was proposed. The proposed LSM algorithm would enable the cloud to predict users' content request distribution with limited information about network and users. The proposed algorithm will also help UAVs choose the optimal resource allocation strategies depending on the network states autonomously. It was shown that the proposed approach yields up to 33.3% and 50.3% gains in terms of the number of users that have stable queues compared to Q-learning with cache and Q-learning without cache. It was also shown that LSM significantly improves the convergence time of up to 33.3% compared to Q-learning.

10.2.4 Ambient Backscatter Networks

To increase SE, a cutting-edge technology named AmBC has received significant attention as a new SS framework [107]. In backscatter communication (e.g. RFID), a device communicates by modulating its reflections of an incident RF signal without generating its own radio waves. Hence, it is in the orders of magnitude more energy-efficient than conventional radio communication. AmBC system enables two devices to communicate using ambient RF as the only source of power. It leverages existing TV and cellular transmissions to eliminate the need for wires and batteries, thus enabling ubiquitous communication where devices can communicate among themselves at unprecedented scales and in locations that were previously inaccessible.

In particular, in an AmBC system as illustrated in Figure 10.8, the backscatter transmitter can transmit data to the backscatter receiver by modulating and reflecting surrounding ambient signals. Hence, the communication in the AmBC system does not require dedicated frequency spectrum. Based on the received signals from the backscatter transmitter and the RF source or carrier emitter, the receiver then can decode and obtain useful information from the transmitter. By separating the carrier emitter and the backscatter receiver, the number of RF components is minimized at backscatter devices and the devices can operate actively, i.e. backscatter transmitters can transmit data without initiation from receivers when they harvest sufficient energy from the RF source [188]. Therefore, AmBC systems can share spectrum with existing systems and achieve better spectral efficiency than that of RFID systems.

The existing ML-based works for AmBC systems are mainly focused on the information extraction and mode selections.

10.2.4.1 Information Extraction

Since ambient backscatter uses uncontrollable RF signals that already have information encoded in them, it needs a different mechanism to extract the backscattered information. Several existing works have proposed different ML-based methods to help extract information.

Figure 10.8 AmBC network. Source: [188] (© [2018] IEEE).

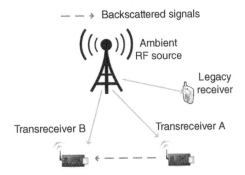

To solve the individual channel estimation in the AmBC system, the authors in [120] designed a communication protocol for the reader and the tag to obtain all the channel parameters. Based on the protocol, they proposed an ML-aided semi-blind estimator which utilizes an Expectation-Maximization (EM) algorithm and a few pilots from RF resources together with some superimposed pilots at the reader. The maximum likelihood estimator was applied to obtain the uplink channel between the reader and the tags, while the superimposed pilots from the reader were used to estimate the downlink channel. Finally, the Cramer–Rao Bounds (CRB) of the proposed channel estimators were derived.

In an AmBC system, the readers receive the backscattered signal from the backscatter device (BD) and the Direct-Link Interference (DLI) from the RF source simultaneously. Due to the randomness of ambient RF sources, it is challenging to distinguish backscatter symbols from DLI. Furthermore, the existence of DLI can further cause the conventional Energy Detector (ED) to fall into severe error-floor problems. To tackle this issue, the authors in [67] developed a novel error-floor free detector by using multiple receive antennas at the reader side. They first considered the perfect CSI case and used beamforming-assisted ED and likelihood ratio-based detector to decode the backscatter symbol. Based on this, a novel statistical clustering framework was designed for joint CSI feature learning and backscatter symbol detection. It was verified that their method can achieve comparable performance with perfect CSI and significantly outperformed the conventional ED.

An ML-assisted AmBC information extraction method was proposed in [200]. The information was modulated on top of the unknown Gaussian-distributed ambient RF signals. The binary phase-shift keying backscatter signals encoded by Hadamard codes can be decoded by the proposed method. By eliminating the direct path signal and correlating the residual signal with the coarse estimate of the ambient signal, the proposed method first extracted the learnable features for the tag signal. k-nearest neighbors' classification algorithm was then employed to recover the tag signals. Finally, a Hadamard decoder was used to retrieve the original information bits from the recovered signals.

The energy detector or Minimum Mean Square Error (MMSE) detector utilized in existing AmBC systems to detect tag signals suffers from a high BER. To overcome this challenge, Support Vector Machine (SVM) and random forest methods were proposed in [80] for detecting the tag signals in an AmBC system by changing the detection problem into a classification problem. To minimize the BER, the proposed method could classify the received signals into different groups based on their energy features.

10.2.4.2 Operating Mode Selection and User Coordination

Due to their passive nature, Backscatter Devices (BDs) in AmBC systems must harvest energy to power operations such as circuit power consumption, transmission, and sensing. BDs need to determine when to switch between communication and energy harvesting modes but the highly dynamic nature and randomness of RF source activities make this switch operation challenging [205]. Moreover, although the BD can perform the backscatter and energy harvesting simultaneously, it is impractical and inefficient when the amount of harvested energy is relatively small and can only supply internal operations. Therefore, how to efficiently determine the tradeoff between energy harvesting and backscattering RF signals is critical in a dynamic environment [189]. Researchers have proposed various solutions based on RL algorithms.

By adaptively selecting the operating mode in a fading channel environment, the throughput maximization problem of the AmBC system was solved in [205]. With the given channel distributions, the problem was modeled as an infinite-horizon Markov Decision Process (MDP) and the optimal mode switching policy was obtained by the value iteration algorithm. When the channel distribution information was unavailable, the Q-learning algorithm was employed to explore a suboptimal strategy through repeated interactions with the environment. The efficacy of their proposed Q-learning method showed that close-to-optimal throughput performance could be achieved.

The authors in [189] proposed an MDP framework to determine the optimal policy for allowing the secondary transmitter to maximize system throughput. The MDP-based optimization requires complete knowledge of environmental parameters such as the probability of a channel state and the successful packet transmission ratio. To cope with these impractical constraints, a low-complexity online RL algorithm was developed that allowed the secondary transmitter to learn from its decisions to discover the optimal policy. To minimize interference, a multicluster AmBC power allocation problem was developed in [87] for short-range information sharing. A Q-learning-based power allocation method was designed to minimize the interference while improving the received SINR. It was shown that the received signal levels could be significantly improved by their proposed scheme.

By considering the strict latency requirements, the authors in [88] employed DQN to solve the communication rate maximization problem for wireless powered ambient backscatter tags. A Q-learning model for ambient backscatter scenarios was developed first, and an algorithm was then proposed that used DNNs to approximate the complex Q-network table.

10.2.4.3 AmBC-CR Methods

Several works combined AmBC with CRN. An RF-powered backscatter CRN enables the secondary transmitter not only to harvest energy from primary signals, but also to backscatter these signals to the secondary receiver for data transmission [72]. Such a combination can provide SUs with potential connection options instead of simply waiting for access opportunities. When the primary channels in RF-powered CRNs that employ AmBC are mostly busy, instead restricting their activity to harvesting energy, the secondary transmitters can use a fraction of the wait time to transmit data by modulating and backscattering the received signals through the AmBC Thus, AmBC enables secondary systems to maximize their performance by simultaneously optimize spectrum usage and energy harvesting.

In an AmBC-assisted CRN network, the mode selection between signal backscatter and energy harvest is critical to achieving high RF-powered SU (RSU) throughput. The dynamics of the primary channel, energy storage capability, and data to be sent all need to be considered when making decision. An MDP-based framework was developed in [84] to determine optimal decisions with consideration to states such as energy, data, and primary channel. It was then expanded to include a scenario in which the state information was unavailable at the RSU. A low complexity online RL algorithm was proposed to enable the RSU to find the optimal solution without requiring prior and complete information from the environment.

10.3 Summary

In this chapter, we have reviewed four SS application scenarios: an opportunistic access-based CRN, a database-assisted SS, an LTE-U/LTE-LAA, and a symbiotic SS mechanism-based AmBC network. The uses of ML in approaching SS related problems with regards to the characteristics of SS frameworks were considered.

11

Secure Spectrum Sharing with Machine Learning: Methodologies

In this chapter, the security concerns of SS will be analyzed, following by some mitigation mechanisms.

11.1 Security Concerns in SS

While the ML-based SS networks can help improve the SE, they can also be a double-edged sword. The dynamic access frameworks introduce more security and privacy risks into the system. As shown in Figure 11.1, when SUs observe the activity of PUs, the sensing procedure can be disturbed by the malicious attackers by launching the PUE attacks or SSDF attacks. The attackers may also exploit these opportunities to harm the privacy of PUs. Besides these, the system also suffers the same security issues found in traditional wireless communications, such as jamming attacks and eavesdropping attacks. In this section, we will discuss these physical layer attacks and potential countermeasures [195].

11.1.1 Primary User Emulation Attack

In CRN, a PUE attack denotes a PU-like signals sent by an attacker during the spectrum sensing period that can exclude legitimate SU access to the channels. The attackers may be selfish users who want to use the spectrum exclusively or malicious attackers who want to disrupt the normal operation of the system. PUE attacks can cause service degradation, denial of service (DoS), connection unreliability, and bandwidth waste [223].

PUE attacks can damage required security such as availability, authentication, non-reputation, compliance, and access control [145]. Countermeasures to PUE attacks seek to enhance spectrum management. When defending against PUE attacks, it is important to differentiate between malicious users and legitimate users. This can be determined by their location, received signal strength, received signal power, and other features.

11.1.2 Spectrum Sensing Data Falsification Attack

CSS as one promising approach for PUs' activities detection involves exploiting the spatial location diversity of multiple SUs. A group of SUs collaborate to perform the spectrum sensing by exchanging locally collected information. An SSDF attack (also known as the

5G and Beyond Wireless Communication Networks, First Edition. Haijian Sun, Rose Qingyang Hu, and Yi Qian.
© 2024 John Wiley & Sons Ltd. Published 2024 by John Wiley & Sons Ltd.

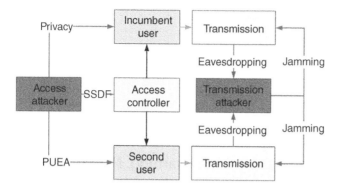

Figure 11.1 Secure issues in SS network.

Byzantine attack) is launched in CSS by sending false local spectrum sensing results to others, leading to flawed spectrum sensing decisions [222]. SSDF attacks aim to decrease detection probability and disturb normal operations of the primary system. It may also seek to increase the probability of false alarms in order to deprive honest SUs of access opportunities [227]. SSDF attacks harm the system's integrity and availability.

SSDF attackers can be classified into three types: selfish SSDF, interference SSDF, and confusing SSDF [222].

(1) A selfish SSDF attacker seeks to gain exclusive access to the target spectrum. It falsely reports the existence of relatively high PU activities to block other SUs from using the spectrum when the PU does not exist.
(2) An interference SSDF attacker falsely reports low PU activities leading other SUs to wrongly conclude that they can use the spectrum without interfering with any PUs. This type of attack seeks to either cause the inference to the PU or inhibit the communication of other SUs.
(3) A confusing SSDF attacker seeks to disturb the SUs to prevent them from reaching consensus by randomly reporting the true or false results about the existence of PUs.

The majority of existing defense methods can be divided into two categories: one making direct judgments based on the current spectrum sensing data while the other uses the historical spectrum sensing data to update sensors' reputation.

11.1.3 Jamming Attacks

The open nature of wireless communication leaves it vulnerable to various attacks. One of the most common attacks in wireless communication as well as SS networks is the jamming attack. Attackers transmit signals to interfere with the victims' communications in order to cause a DoS and compromise availability of communication links [104]. Traditional anti-jamming methods used in wireless communications include sequence-based frequency hopping spread spectrum (FHSS) and direct sequence spread spectrum (DSSS). However, the fixed transmission patterns of these methods leave them helpless against dynamic jamming attacks and cause low spectrum efficiency.

SS techniques enable flexible access to different channels, allowing users to avoid attackers by exploiting that flexibility. The ML techniques provide more adaptive channel selection ability to systems in order to avoid jamming attacks. They also give the system the ability to learn and predict the behavior of jammers to increase anti-jamming channel selection efficiency. The attackers may also use different ML-based methods to improve their attack strategies [206] rendering the study of advanced jamming attacks and corresponding countermeasures of vital importance to the SS system.

11.1.4 Intercept/Eavesdrop

Eavesdropping is another common attack in wireless communications. Due to the broadcast nature of radio propagation, any active transmissions operated over the shared spectrum by different wireless networks are extremely vulnerable to eavesdropping. It is therefore important to investigate the confidentiality protection of SS communications against eavesdropping attacks [242].

There are two major categories of secure communication techniques that guard against eavesdropping. One focuses on traditional cryptographic techniques and the other is the physical layer security. Cryptographic techniques involve encryption and decryption of information at the transmitter and receiver. In the physical layer security method, the secrecy rate can be achieved by the mutual information difference between the legitimate user and the eavesdropper. However, the security rate can be limited since it depends on the difference between the channel condition from the transmitter to the legitimate receiver and that from the transmitter to the eavesdroppers. Many promising techniques have been proposed to address this issue, including artificial noise (AN) and cooperative jammer (CJ) [203]. The advantage of physical layer security over cryptographic is that it can achieve secure communications without extra overhead caused by protecting the security key and can therefore be used in relatively simple communication systems.

11.1.5 Privacy Issues in Database-Assisted SS Systems

According to [121], there are several differences between security issues and privacy issues. Security issues refer to unauthorized/malicious access, change, or denial of data. Privacy issues refer to the unintentional disclosure of sensitive information from some open-access data. The former is usually the work of malicious attackers who wish to disturb the system. In the latter, malicious users usually only collect information that does not immediately cause direct harm to the system. The goals of security protection are confidentiality, integrity, and availability. The goals of privacy protection are anonymity, unlinkability, and unobservability.

Ensuring privacy in SS networks is very important. In some SS systems like SAS in the CBRS band, the IUs can be radar devices and military ships carrying very sensitive information. Opening these incumbents' exclusive spectrum to sharing could usher in potential privacy threats to the system. Furthermore, the distribution structure of the SS network increases the risk of privacy leakage. Finally, in order to train itself, ML requires huge amounts of data that may contain various private user information that must be protected during training and communication.

One possible attack is the database inference attack (DIA), where malicious users can obtain PU location and other private information through collected data and sophisticated inference techniques. In another form of attack, the operational privacy threat of SUs comes from the untrustworthy database that collects the location information sent from SUs on the set of available channels in their region.

The protection of PUs privacy cannot be addressed by strictly controlling access to the database, since each SU must access it to enable the spectrum sharing process. One possible solution might be to reveal obfuscated information instead of the original information to SU queries. By doing this, the system can use the obfuscated information to help determine the channel status while reducing leakage of PU's privacy information. SUs' privacy can also be protected by sending an obfuscated version of the original information of SUs. Some works also looked at the question of how much information to share with the database and the dynamic question of whether to share information with the database.

ML algorithms require massive amount of data to train their models. These data usually include a lot of user-specific sensitive info and need to be exchanged in some distributed systems. Sensitive information may leak out during the training process that would have remained secure using the above spectrum sensing procedure. Three main strategies may be used to maintain privacy in ML work flow: differential privacy, homomorphic encryption, and Secure Function Evaluation (SFE)/Secure Multi-party Computation (SMC) [102]. In the differential privacy method, publicly shared dataset information describes the patterns of groups within the dataset but withholds information about individuals. In homomorphic encryption, the operation on encrypted data can be used to secure the learning process by computing on encrypted data. When user-generated data are distributed among different data owners, SFE can enable multiple parties to collaboratively compute an agreed-upon function without leaking input information regarding any party other than what can be inferred from the output.

11.2 ML-Assisted Secure SS

In this section, the latest research in ML-related security will be comprehensively surveyed. The contents are organized as in Figure 11.2. Details of existing works will be reviewed in each category.

11.2.1 State-of-the-Art Methods of Defense Against PUE Attack

In this section, different ML-based PUE attacker detection methods in [4, 12, 50, 52, 60, 86] will first be presented. To address the limitations of training data and inconsistent communication environments, the robust detection methods in [35, 125, 162, 168] will be further discussed. Finally, the ML-based attack strategies included in [37, 38, 156] will be discussed to provide an attacker's perspective on PUE defense method design.

11.2.1.1 ML-Based Detection Methods

A typical PUE attack is illustrated in Figure 11.3. In defending against such attacks, the most important step is distinguishing malicious attackers from legitimate PUs. This can be achieved using specific features extracted from received signals. Distinct features may reflect the transmitters' characters, rendering them unique and differentiable.

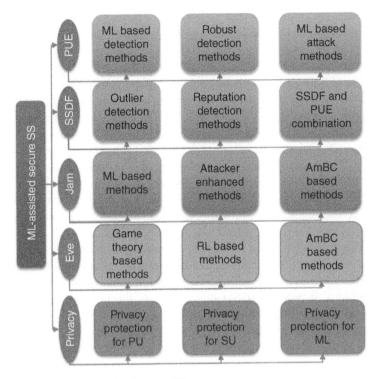

Figure 11.2 ML-assisted secure SS.

Figure 11.3 Illustration of PUE attacks.

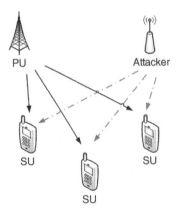

User location-based method is a common and easy way to differentiate between attackers and PUs. Malicious attackers are rarely in the same place as PUs. Since the received signal strength (RSS) varies by location, it can be adopted to identify location and, by the same token, user type. Some other methods are based on statistical analysis. They use features such as signal power, spectrum occupancy time, and cyclostationarity extracted from received signals to analyze transmitters. Finally, the physical layer approaches use the hardware behaviors of transmitters or channel behaviors to detect attackers. For example, phase and frequency shifts are commonly used as transmitter fingerprints. A detection problem based on received signals is a classification problem, which ML is particularly good at solving.

To detect and defend against PUE attacks, an adaptive learning-based detection method in CRN was developed in [12] by analyzing the transmitters' power features. Specifically, cyclostationary feature analysis was used to differentiate attackers from low-power PUs. The proposed method also estimated the distance variance and communication time to improve classification accuracy and communication rate.

A k-nearest neighbor (KNN) classifier-based detection method was used to classify malicious users to forestall PUE attacks in [86]. KNN was trained with the parameters such as data rate, distance, power, frequency of request. Moreover, Elliptical Curve Cryptography (ECC) was applied to encrypt the data and improve network security. the proposed classifier achieved a higher accuracy detection performance than the Artificial Neural Network (ANN)-based method.

A channel-based method that relied on the behavior of the multi-path channel was investigated in [4], where the authors proposed an ML framework based on various classification models for detecting PUE attacks. It was trained/tested using four features vectors extracted by the Pattern Described Link-Signature (PDLS) method. By using this method, legitimate and malicious users could be effectively distinguished.

When signal activity patterns can be regarded as a possible sequence relating to some "features" of the channel and previous internal states, the Recurrent Neural Network (RNN) is a good tool for PUE detection. An RNN-based PUE attack detection method was first introduced in [50]. It exploited series' temporal dependency for better series prediction and abnormal activity detection. To deal with the gradient vanishing issues inherent to RNN, an advanced version of RNN that took advantage of the Long Short Term Memory (LSTM) units and processed time series with long-term memory more efficiently was further proposed. It was shown that the LSTM-based method could significantly improve the detector's performance.

The authors in [60] investigated the joint detection of PUE and jamming attacks in CRN. A sparse coding of the compressed received signal-based detection algorithm was proposed. Based on the channel dependent dictionary, convergence patterns in sparse coding were employed to differentiate the spectrum hole, legitimate PU, and emulators or jammers. An ML-based classification was adopted to perform the decision-making operation. The effectiveness and advantages of the proposed algorithm were verified in terms of the confusion matrix quality metric.

By detecting PUE attacks and enhancing the probability of detection, a hybrid Genetic Artificial Bee Colony (GABC) algorithm was proposed in [52] to optimize spectrum utilization. A Genetic Algorithm was used to compensate for the Artificial Bee Colony algorithm's less than optimal exploitation of solutions by using crossover and mutation operations. The proposed GABC incorporated the Genetic operators into the Artificial Bee Colony algorithm to achieve balance between exploitation and exploration in order to find the optimal solution.

11.2.1.2 Robust Detection Methods

Most of the classification approaches discussed above require a certain dataset to train the network. However, an unfamiliar environment and attackers may lead to classifier failure due to an unsuitable reference model. Hence, robust detection methods are very important when adapting to the changes in the environment and more practicable in the real world.

In [35], defense strategies against PUE attacks from malicious SUs were investigated. Greedy SUs manipulated unsupervised learning clustering algorithms through various attack strategies to evict other SUs from the idle channel. Corresponding countermeasures to these manipulation attacks were developed, and the robustness of unsupervised learning for signal classification in an evolving RF environment was evaluated. By using both k-means clustering and Self-Organizing Maps (SOMs), the proposed method could perform signal classification in the absence of training data. With intentional attacks from SUs, the robustness of the classifier to avoid misclassification was verified. It was shown that the efficacy of attacks could be reduced by 75.9%.

By adopting the TL algorithms, the authors in [162] developed a PUE defense approach that used knowledge about PUs and SUs from past time frames to improve the detection process in future time frames. The proposed approach extracted high-level representations of the environment and accumulated them to form an abstract knowledge database. This database enabled the CR system to accurately detect PUE attacks even if an insufficient amount of fingerprint data was available in the current time frame. The final detection decisions were used to update the abstract knowledge database for future runs.

A semi-supervised distributed learning algorithm was proposed for PUE attack detection. By enabling edge devices to perform data clustering and session classification locally, it could deal efficiently with varying bandwidth, signature changes, etc. The labeled data was fed into a trained supervised learning-based classifier for classification. Based on the error rate, it adjusted the training vectors and improved the overall performance. It was shown that the proposed method could significantly reduce false alarms in the secondary network and improve overall detection accuracy in the primary network.

In [125], an adaptive Bayesian learning automaton algorithm-based scheme named Multi-channel Bayesian Learning Automata (MBLA) was proposed to defend against PUE attackers. The SU in the considered model adopted Uncoordinated Frequency Hopping (UFH) to avoid PUE attacks. To improve the speed and accuracy of the learning process in non-stationary environments, MBLA utilized two different channels simultaneously to perform the optimal frequency channel selections. Statistical information about channels and PUs was assumed unavailable. An SU synchronized with its receiver and sent its data on various channels obtained by the MBLA. The scheme extracted the best strategies for the attacker and the SU and then evaluated the proposed scheme in terms of the SU throughput in the presence of the PUE attacker.

11.2.1.3 ML-Based Attack Methods

Throughout the above discussions, various PUE attack detection and defense strategies have been offered. However, the best-attack strategies have not been discussed. A better understanding of optimal attack strategies can enable researchers to quantify the severity or impact of a PUE attacker on an SU's throughput. It can also shed light on practical defense strategy design as the attackers can also exploit ML algorithms to improve their performance.

In [156], the authors presented two GAN-based models that successfully emulated the PUs. Depending on whether any prior knowledge of the PU's feature space was available, they proposed a dumb generator model and a smart generator model. Two DNN-based discriminator models were developed to distinguish the PU and the Emulated PU (EPU) from the corresponding generators. With iterative and sequential training, the generator and discriminator of each GAN model became smarter and smarter. It was shown that discriminators were able to detect about 50% of PUE attackers without the GAN training during the deployment phase, and both the GAN models could achieve 100% accuracy during the training phase. After the GAN training, the discriminators of the dumb generator could achieve 98% accuracy while the smart generator-based model could achieve 99.5% accuracy.

Optimal PUE attack strategies were investigated in [37], where prior knowledge on PU activity characteristics and SU access strategies was not available. Based on previous attacking experience, a non-stochastic online learning problem was formulated to determine attacking channel decisions for attackers. Since a PUE attacker never knows if an SU has ever tried to access the attacked channel or not, it cannot observe the reward. Therefore, an Attack-But-Observe-Another (ABOA) scheme was proposed to solve this issue. The attackers in this scheme attack one channel in the spectrum sensing phase but observe one or more other channels in the data transmission phase. Two non-stochastic online learning-based attacking algorithms, EXP3-DO and OPT-RO, were proposed to select the observing channel deterministically based on the attacking channel and uniform randomly, respectively.

11.2.2 State-of-the-Art Methods of Defense Against SSDF Attack

CSS can help overcome the fading environments and improve the system sensing performance. Different from single-user-based SS, each SU needs to transmit the sensing results to a Fusion Center (FC) in CSS. FC then combines those results and makes a final decision about the PU's presence. SSDF is the most common attack in CSS. As shown in Figure 11.4, sending falsified sensing data to the FC can lead to an incorrect fusion result, cause interference with PUs, and cause DoS to SUs. To defend against SSDF attacks, the most important step is to differentiate attackers from legal SUs. The existing defense methods fall into two groups, namely outlier detection approaches and reputation-based approaches. In outlier detection methods, the abnormal user is excluded from the network. In reputation-based methods, on the other hand, SUs are assigned a reputation degree that

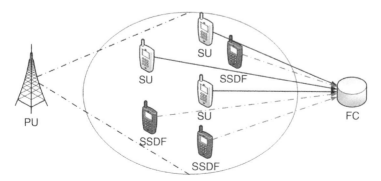

Figure 11.4 Illustration of SSDF attack.

reflects their detection performance. Since SUs are not eliminated and their reports are not excluded, reputation-based methods can use the collected information more thoroughly than outlier detection techniques.

11.2.2.1 Outlier Detection Methods

To mitigate SSDF attacks, the authors in [55] proposed a Support Vector Data Description (SVDD) algorithm in the sensing phase that could distinguish malicious nodes from legal ones and remove them in the decision phase. The boundary around the target data was constructed by enclosing the target data within a minimum hyper-sphere. Enlightened by the SVM, the SVDD decision boundary was described by a few support vectors. The spectrum sensing result was then decided according to the voting results from trusted nodes.

In [1], the authors designed a Bayesian nonparametric clustering scheme to sense the primary channel status and identify malicious users in CSS. By forming a single cluster consisting of all legal users and a separate cluster that included every variety of attacker (selfish, mischievous, jamming) in the feature space, their proposed approach could detect and identify multiple attacks simultaneously. Furthermore, based on observations from the collaborating CR users, it could discern malicious users from the legitimate ones and obtain the real PU traffic pattern.

In [139], Bayesian learning-based SSDF defense schemes were proposed. First, a Bayesian offline learning algorithm was proposed where the spectrum state was unavailable for training. A Bayesian online learning framework was then designed by incorporating the time-varying attributes of the sensors. The framework consisted of the historical data learning part and the current data learning part. The vector of sensors' weight was updated by considering both the historical and the current spectrum sensing data. Finally, an SSDF attack behavior recognition algorithm based on the proposed framework was designed to identify SSDF attacks more accurately than offline learning.

11.2.2.2 Reputation-Based Detection Methods

One of the critical issues in combating SSDF attacks is distinguishing the attackers' error reports from the SUs' reports in the FC. A Bayesian reputation model-based SSDF defense scheme for CRNs was proposed in [131]. The proposed method modeled cooperation as a service-evaluation process and SUs' reputation degrees reflected their service quality. Reputation degrees of SUs were updated based on the Bayesian reputation model, and the behaviors of malicious SUs could be effectively tracked by this means.

In [83], a three-layer Bayesian model was designed to combat SSDF attacks. The model consisted of a processing layer, an integrating layer, and an inferring layer. The processing layer was based on the HMM model, where original data was used to train parameters and the trained emission distributions were then passed to the second layer. By employing different algorithms in the integrating layer, emission distributions were processed to obtain reputation values, balance values, and specificity values of different SUs. By using different thresholds, these continuous values could be rendered discrete and then transferred to the inferring layer. Finally, by using the discrete values as evidence, a Bayesian network was built in the third layer to calculate the safety probabilities of SUs.

In [160], several ML techniques such as SVM, Neural Network, Naive Bayes, and Ensemble classifiers were implemented to detect SSDF attacks in a CRN. The learning techniques were investigated under two experimental scenarios: (i) the training and test data were drawn from the same data-set, and (ii) separate datasets were used for training and testing. The robustness of the proposed ensemble method was verified compared with other benchmarks.

An SVM-based scheme was proposed in [240] to deal with SSDF attacks. SUs' behaviors were analyzed from multi-round records of energy values, and their classification accuracy was obtained. Furthermore, the concepts of recognition probability and misclassification probability were introduced, and the tradeoff relationship between misclassification probability and threshold of classification accuracy was obtained. As a result, the proposed scheme enabled excellent adaptability for Malicious SU (MSU) detection in various scenarios.

11.2.2.3 SSDF and PUE Combination Attacks

The combination of the SSDF attack and the PUE attack presents more challenges to the network. If SUs are be attacked or mislead by the PUE attack, the performance of existing SSDF defense methods is degraded. Even if the PUE attacker is detected, neighboring SUs can still submit flawed sensing reports due to the contaminated signal from the PUE attack. Investigating the combination of attacks and corresponding ML-based countermeasures in existing works can provide a comprehensive understanding of secure design in SS networks.

Secure sensing under both PUE and SSDF attacks in CRN was investigated in [99]. Directly excluding attacked SUs from the sensing cooperation process requires lots of information, i.e. the attack strength, geographical locations of SUs, etc. Therefore, a novel secure sensing algorithm was developed to deal with the problem. To be specific, Unsupervised ML (UML) was adopted to identify contaminated sensing reports from trusted users by examining their sensing history. These contaminated sensing reports were then excluded from CSS. Moreover, considering that identification errors might occur during the UML process, each SU was assigned an identity value to account for its reliability. The identity value was also used to alleviate the misidentification impact on real trusted users.

To defend against various malicious attacks and interference in full-duplex CRNs (FD-CRNs), an ensemble ML (EML)-based robust CSS framework was proposed in [231, 232]. SUs were assumed to have the ability to sense and transmit over the same frequency band simultaneously. The self-interference and co-channel interference were inevitably introduced into the system and complicated the sensing environment. By investigating spectrum waste probability, collision probability, and secondary throughput in both FD LBT and Listen-and-talk protocols, the robust and accurate fusion performance of the proposed EML approach was verified.

11.2.3 State-of-the-Art Methods of Defense Against Jamming Attacks

A jamming attack is a common in wireless communication systems and there have been many works on this subject. Due to the nature of SS, jamming attack is effective, hence can cause severe damages. In the following, we review latest advances in anti-jamming with ML.

11.2.3.1 ML-Based Anti-Jamming Methods

In the SS network, except for the very rapid change of dynamic spectrum characteristics in the channel, the inclusion of random jammers makes efficient communication more challenging.

To defend against jamming attacks, the most common countermeasure strategies are dynamic channel assignments based on different ML algorithms. The jamming attack scenario can be modeled by using the stochastic zero-sum game and MDP framework. The time-varying characteristics of the channel as well as the jammer's random strategy can be learned by the SU using RL algorithms.

A stochastic game framework was proposed in [193] for anti-jamming defense. At each stage of the game, SUs observe the spectrum availability, the channel quality, and the attackers' strategy from the status of jammed channels. Based on observation results, the number of reserved channels and the channel switch action policy are decided. SUs employ Minimax-Q learning to learn the optimal policy, maximizing the expected sum of discounted spectrum-efficient throughput. It was shown the proposed stationary policy in the anti-jamming game performed better than the myopic learning and random defense strategy because it successfully accommodated the environment dynamics and strategic behavior of the cognitive attackers. To further improve system performance, the authors used the QV and the SARSA RL algorithms in [164] to replace the Minimax-Q learning in [193]. Minimax-Q learning is an off-policy and greedy algorithm, whereas the QV and SARSA are on-policy algorithms. It was shown that QV learning can achieve the best performance as the value of both Q and V is updated.

In [206], the authors first investigated an anti-jamming game model where the SU could access only one channel at a time and hopped among different channels. An MDP-based channel hopping defense strategy with the assumption of perfect knowledge was derived by analyzing interactions between the SU and attackers. Based on this, they proposed two learning schemes by which SUs gained knowledge of adversaries in order to handle cases without perfect knowledge. The schemes were then extended to a scenario where SUs could access all available channels simultaneously and redefined the anti-jamming game with randomized power allocation as the defense strategy. The Nash equilibrium was derived for this Colonel Blotto game, which minimized the worst-case damage.

In [69], a game model was formed to integrate anti-jamming and jamming subgames into a stochastic framework. Q-learning was applied to find an optimal channel access strategy. It was shown that Minimax-Q learning was more suitable than Nash-Q learning for an aggressive environment. For distributed mobile ad hoc networking scenarios, Friend-or-foe Q-learning provided the best solution where centralized control was nearly unavailable.

By employing the Double Q-learning algorithm to defend against the jamming attacks, Multi-Objective Ant Colony Optimization (MOACO) and greedy-based optimization methods were proposed in [192]. A Q-learning-assisted cluster-based data utilization was proposed that could enhance inter-cluster data aggregation. The network lifetime was improved using AI-based modeling with intra-network to enhance green communication. Unlike the artificial bee colony and genetic algorithm, the throughput, device lifetime, and jamming prediction were promoted using the proposed MOACO.

A Wideband Autonomous CR (WACR) anti-jamming method presented in [123] evaded a jammer that swept across the whole wideband spectrum range. The WACR equipped spectrum knowledge acquisition ability to detect and identify the location of the sweeping jammer. A Q-learning-based method was proposed to allow the anti-jamming operation to cover over several hundred MHz of a wide spectrum in real-time. An anti-jamming-based secure communication protocol was then developed that selected a spectrum position with enough contiguous idle spectrum to resist interference by both deliberate jammers and inadvertent disruptions. The communication then switched to this position until the jammer arrived. When the jammer began to interfere with the CR's transmission, it switched to a new spectrum band that led to the longest possible uninterrupted transmission as learned through Q-learning. By including more agents in the system, the authors in [11] further proposed an advanced RL-based anti-jamming approach. The considered system model allowed multiple WACRs to operate over the same spectrum band simultaneously. Each radio attempted to evade other WACRs' transmissions and avoid jammer signals that swept across the whole spectrum band of interest. The WACR first detected and identified the frequency location of this sweeping jammer and the signals of other WACRs. A sub-band selection policy was then given by the RL-based approach based on the detection results to avoid both the jammer signal and interference from other radios.

To enable network devices to detect and predict jamming signals in a system with multiple jamming modes and noises, it is critical to develop a rapid jamming detection countermeasure. To this end, the authors in [27] proposed a DL-based jamming pattern recognition by using spectrum waterfall. In addition, the simplified Le-Net5 structure was employed to reduce the complexity of the calculation. As a result, the proposed method achieved a rapid recognition performance.

By directly using temporal and spectral information like spectrum waterfall, the authors in [114] developed a novel anti-jamming approach that did not require knowledge of jamming patterns and parameters. First, a recursive CNN was designed to overcome the infinite-state issues of spectrum waterfall. Furthermore, a DRL algorithm-based anti-jamming method relying only on locally observed information was proposed to obtain optimal anti-jamming strategies. The proposed method could explore the unknown environment and combat advanced jamming attacks in a more practical fashion.

A sequential DRL algorithm without prior information was proposed in [116] to defend against jamming attacks. The jamming patterns were first identified by DL and sliding window principles. Those recognized patterns were then passed to an RL-based model to inform online channel selection. To better achieve the tradeoff between throughput and overhead, channel switching cost was introduced to the system. It was shown that the proposed method could make anti-jamming channel selection decisions quickly without modeling the jammer's characteristics.

11.2.3.2 Attacker Enhanced Anti-Jamming Methods

Although ML provides many effective solutions for system defense when fighting jamming attacks, it can also be exploited by attackers to develop more effective attack strategies. Considering intelligent attackers when designing defensive measures can help avoid overly clumsy assumptions and enhance protection schemes more reliable and practical.

In addition to anti-jamming techniques, knowing intelligent jamming strategies is also crucial. An intelligent jammer that can adapt to its surroundings was investigated in [7] under an electronic warfare-type scenario. To be more practical, the delay of packet exchange information between the victim senders and the receivers was considered by the jammer, as opposed to the traditional assumption where the feedback is instantaneously available. Furthermore, to implement delayed learning in scenarios where rewards were associated with state transitions, a new method was developed. The advanced benefits of the proposed framework were verified by studying the optimal jamming strategies against an 802.11-type wireless network that used the RTS-CTS protocol to communicate and deliver information.

To jam the SU communications without interfering with the PUs, a cognitive jammer with sensing capability can exploit the same statistic information and stochastic dynamic decision-making process that SUs would follow. To this end, an anti-jamming multi-channel access problem was formulated in [196] as a non-stochastic multi-armed bandit problem. By taking advantage of shared information among the transceivers, a protocol was developed that enabled SUs to selectively sense channels with a high probability of non-occupancy by jammers and PUs based on the sensing and access historical information.

To proactively avoid jammed channels, Q-learning was employed to learn strategies of jammers in [165, 166]. Due to the time-consuming training process required by Q learning for learning the behaviors of jammers, a wideband spectrum sensing ability was adopted to speed up the learning process. Prior learned information was also used to minimize the number of collisions with the jammer in the training phase. Finally, the effectiveness and improvement of the modified algorithm were verified.

As shown in Figure 11.5, an adversarial ML approach launching jamming attacks and introducing a defense strategy was presented in [157, 163]. A transmitter T first sensed channels and identified spectrum opportunities, then transmitted data in idle channels.

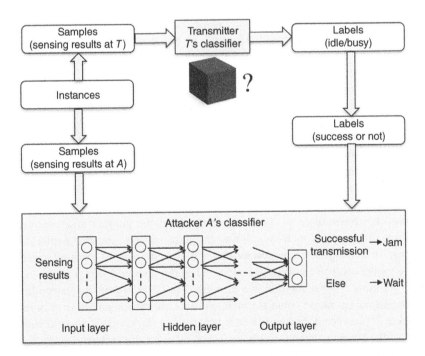

Figure 11.5 System model for attacker's learning. Source: [163] (© [2018] IEEE).

In the meantime, an attacker A also sensed channels and identified busy channels with the intention of jamming legitimate users' transmissions. A pre-trained ML algorithm was implemented at T to classify a channel as idle or busy. This classifier was unknown to the attacker, while A also sensed the channel to capture T's decisions by tracking the acknowledgments. By applying a DL with inference attack, the attacker also built a classifier that was functionally equivalent to the one at the transmitter. Therefore, A could reliably predict successful transmissions based on the sensing results and effectively jam these transmissions. By exploiting the sensitivity of DL to training errors, a defense scheme was then developed by T to defend adversarial DL. The transmitter deliberately used a small number of wrong actions to launch a poisoning attack on the attacker when it accesses the spectrum. The goal is to prevent A from building a reliable classifier. To this end, T systematically decided when to take wrong actions to balance the conflicting effects of deceiving A and making correct transmission decisions. This defense scheme successfully fools the attacker into making prediction errors and allows the transmitter to sustain its performance under intelligent jamming attacks.

11.2.3.3 AmBC Empowered Anti-Jamming Methods

Users in the AmBC network are vulnerable to interference and jamming since their operations are based on ambient RF signals with a limited power supply. However, every cloud has a silver lining. The jamming attacks can be used as additional sources of energy and information by AmBC empowered systems.

To observe the performance of the AmBC system under a jamming attack, the interaction between a user and an intelligent jammer was modeled as a game in [151]. The backscattering time utility functions of both user and jammer were designed, and the closed-form expression for the equilibrium of the Stackelberg game was obtained. As the system SNR information and transmission strategy of the jammer were not available, Q learning was employed to obtain the optimal strategy in a dynamic iterative manner. Hot booting Q-learning was further introduced to accelerate the convergence of traditional Q learning.

Most jamming countermeasures focus on how to enable users to efficiently escape the invaded channel. AmBC opens the possibility of fighting against the malicious jammer. As shown in Figure 11.6, a method that allowed wireless nodes to fight against a jamming attack instead of escaping was proposed in [85, 190]. By first learning the adversary's jamming strategy, the users could decide whether or not to adopt the rate or backscatter modulated information on the jamming signals. A dueling neural network architecture-based DRL algorithm was developed to deal with unknown jamming attacks such as jamming strategies, jamming power levels, and jamming capability. The proposed algorithm allowed the transmitter to effectively learn about the jammer and conceive optimal countermeasure actions such as adapting the transmission rate, backscattering, harvesting energy, or staying idle. The system performance in terms of learning speed, throughput, and packet loss was significantly improved by the proposed algorithm.

An RL-based jamming defense method was developed in [191], where the transmitter could obtain the optimal operation policy through real-time interaction processes with the malicious attacker. To be specific, when the jammer attacked the channel, the transmitter could leverage the jamming signals to transmit data by using the ambient backscatter technique or harvest energy from the jamming signals to support its operation. Thus, the

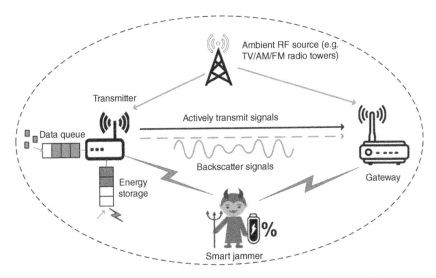

Figure 11.6 Anti-jamming attack in AmBC-CRN. Source: [190] (© [2019] IEEE).

proposed method enabled the transmitter to transmit data even under jamming attacks. It was also observed that the more power the jammer used to attack the channel, the better the network performed.

11.2.4 State-of-the-Art Methods of Defense Against Intercept/Eavesdrop

In eavesdropping attacks, an attacker tries to intercept private information from the legalized communication system. The basic principle of all defense methods is keeping leakage of information at an acceptable level. The encryption methods aim to totally block this leakage, while the physical layer security methods maintain the leakage rate under the required threshold by using different methods such as providing a higher channel difference, introducing friendly jammers, and/or adding artificial noise. Many ML-based works have been proposed to defend against eavesdropping attacks.

11.2.4.1 RL-Based Anti-Eavesdropping Methods

A multilevel Stackelberg game-based secrecy transmission of CRN under an eavesdropper attack was considered in [169]. To protect the achievable rate, some SUs acted as the trusted decode and forward relays. Moreover, to proactively protect the legitimated transceivers, some SUs offered friendly jamming services and requested corresponding service charge prices. Furthermore, an advanced encryption method was adopted to increase the effective security level when users accessed the primary spectrum in the presence of eavesdroppers. By this means, the achievable rate was maximized and the consumed power minimized. Finally, a fuzzy-based MDP Outcome Prediction (MDPOP) Q learning algorithm was proposed to eradicate eavesdropping occurrence in CRNs.

In [82], a DRL-based relay selection for secure buffer-aided CRNs was investigated. Considering that an eavesdropper keeps intercepting the signals from the source and relays, the relay selection problem was modeled as an MDP problem to protect the transmission data. A DQN-based approach was introduced to solve this MDP problem, and the ϵ-greedy strategy was applied to balance the exploitation and exploration.

A secure EE-based communications problem with energy harvesting ability in CRN was investigated in [48]. With the limited energy supply and presence of passive eavesdroppers, a TL actor-critic learning-based algorithm was introduced to help the SUs determine their operation mode to achieve a higher security level. In particular, SUs interact with the environment directly and choose to either stay idle to save energy or transmit the encrypted sensing results to FC by using a suitable private-key encryption method to maximize the long-term effective security level of the network.

In real applications, some government agencies need to locate suspicious communications via legitimate eavesdropping in an efficient manner. To this end, it is necessary to study the optimal attack strategies for energy-constrained eavesdroppers. A full-duplex active eavesdropper with a limited energy budget was considered in [184]. It sought to capture data and interfere with suspicious transmission links. A legitimate attack optimization problem was formulated based on a partially observable MDP framework to maximize the achievable wiretap rate while minimizing the suspicious throughput over a Rayleigh fading channel. Based on the available energy and beliefs regarding licensed channel activity, eavesdroppers needed to determine the course of action with maximum long-term system benefits. This may be either passive eavesdropping without jamming or active eavesdropping with an optimal amount of jamming energy.

11.2.5 State-of-the-Art ML-Based Privacy Protection Methods

In this section, privacy protections for PU, SU, and ML in different SS frameworks are sequentially reviewed.

11.2.5.1 Privacy Protection for PUs in SS Networks

In some SS networks such as SAS and LSA, leakage of PU's privacy can cause serious security threats. Potential malicious adversaries may exploit attacking technologies such as DIA to obtain information about the IU and that can later be used to develop attack strategies. To defend the DIA and protect IUs, one viable approach is to obfuscate the information revealed by the database. There are a lot of works using ML-based obfuscation techniques to counter inference attacks.

The authors in [43] investigated whether or not a malicious opponent could infer the radar's location through veiled radar information contained in the system's precoder. An ML-based location inference attack was simulated for two specific precoder schemes. It was verified that radar privacy could be compromised by system information leaked through the precoder, introducing various degrees of risks.

The tradeoff between privacy preservation and spectrum efficiency was examined in [36]. A generalized SS system architecture was proposed and a multi-utility user privacy optimization problem was formulated. Potential adversary inference attacks were used to measure privacy, and an efficient heuristic strategy was developed to solve the problem. Compared with existing obfuscation strategies, the proposed approach can achieve a 50% increase in privacy with an insignificant impact on SE.

In [187], the authors first showed that adversarial SUs could employ a Bayesian learning-based inference algorithm to accurately locate a non-stationary radar system using only information gathered from seemingly innocuous query replies obtained from a SAS. Several obfuscation techniques were then proposed and implemented in the SAS for countering such inference attacks. Finally, the obfuscation techniques' efficacy in minimizing spectral efficiency loss while preserving incumbent privacy was investigated.

The authors in [202] proposed a CS-based federated learning framework to achieve IU detection for improving communication efficiency while protecting the privacy of training samples. By using an MMV CS model, each sensor transmitted the updated aggregated parameters instead of the raw spectrum data to the central server to protect privacy. They demonstrated that the detection performance was as good as the scheme under the raw training samples, while significantly improving the communication and training efficiency.

11.2.5.2 Privacy Protection for SUs in SS Networks

In [61], several location privacy-related attacks in CSS-based CRNs were first identified. Such attacks can threaten SUs' location privacy by correlating their sensing reports and their physical location. To prevent leakage of location privacy, a privacy-preserving framework was proposed. It was demonstrated that without efficient protection, the attackers could compromise a SU's location privacy at a success rate of more than 90%. A proposed privacy-preserving framework was further introduced and verified, which could significantly improve the location privacy of SUs with a minimal effect on the performance of collaborative sensing.

To protect the location privacy of SUs while allowing them to sense vicinity spectrum availability, two location privacy-preserving schemes for database-driven CRNs were studied in [66]. The spectrum databases' structured nature and SUs' queries were exploited by those schemes to create a compact representation of databases that could be queried by SUs without requiring them to share their location with the database, thereby eliminating the possibility of location leakage. Based on whether a user is a member of a set or not, the first method, location privacy in database-driven CRNs (LPDB), constructed a compact version of the database and provided optimal location privacy to SUs in the coverage area. It achieved unconditional security with an acceptable communication overhead. The second method, LPDB with two servers (LPDBQS), minimized SUs' overhead with an additional network entity cost. The tradeoff between cost and performance provided more options for system design based on specific requirements.

In [237], an aggregative game was used to model SS in large-scale, heterogeneous, and dynamic network. By utilizing past channel access experience, an online learning algorithm was proposed to improve the utility of each user. Considering the heterogeneous impact of users, a multi-dimensional aggregative game was used to model the SS of the large-scale wireless network. A mediated privacy-preserving and truthful mechanism was developed to achieve an η-approximate ex-post NE and provided no regret guarantee for each user. It was shown that the proposed method satisfied joint differential privacy.

11.2.5.3 Privacy Protection for ML Algorithms

The sensitive training data in ML-based applications faced distinct privacy issues. Malicious attackers can obtain private information through the structure of models or their observations. To investigate the privacy leakage of training data, the authors in [221] introduced novel formal definitions of advantage for membership and attribute inference attacks. Attacks in different learning algorithms and model properties were analyzed based on these definitions. It was shown that overfit could increase the risk of privacy leakages.

Secure Multi Party Computation (MPC) allows different entities to share joint data to train their models without releasing any private information in the training data. The MPC-based privacy-preserving method was investigated in [134] for linear regression, logistic regression, and neural network training using the stochastic gradient descent method. A two-server model was considered, and the training data was securely distributed among two non-colluding servers. Different models were trained on the joint data using secure Two Party Computation (2PC). It was shown that their new techniques could significantly increase speed while guaranteeing performance without leaking data privacy.

Cloud computing frameworks provide many benefits to the communication networks, such as powerful processability and unlimited storage space. However, some cloud services are provided by third parties such as Amazon AWS, Google Cloud, Microsoft Azure, etc. Users may hesitate to entrust their sensitive data to these entities, and rightly so.

To protect different data owners' privacy, the authors in [70] introduced a new efficient method that allowed all participants to publicly verify the veracity of the encrypted data. A Unidirectional Proxy Re-Encryption (UPRE) method was also adopted to lower the computation costs. A noise was further added to the encrypted data to preserve the private information while guaranteeing the effectiveness of ML training on cloud.

A cloud-assisted privacy-preserving ML framework was developed in [224]. By using outsourced ML algorithms, the cloud server first generated a model, then processed testing data from the network with the generated model in real-time. The proposed framework adopted a differential privacy method of performing privacy-preserving data analysis and homomorphic encryption in order to conduct valid operations over encrypted data.

FL allowed decoupling of data provision and ML model aggregation and shows promise as a framework for addressing privacy problems for distributed ML [178]. It enables the users to cooperatively learn a global model without sacrificing data privacy directly. The information transmitted for FL consists of minimal updates to improve a particular machine learning model. However, the design of FL still needs the protection of parameters as well as investigations on the tradeoffs between the privacy-security-level and the system performance. The study [62] suggested that FL could expose intermediate results such as stochastic gradient descent, and the transmission of these gradients may actually leak private information when exposed together with a data structure. It is still possible for adversaries to reconstruct the raw data approximately, especially when the architecture and parameters are not completely protected.

To investigate the leakage of private information in users' data, the performance of malicious servers was studied in [201]. A GAN-based framework with a multi-task discriminator capable of discriminating category, reality, and client identity included in input samples simultaneously was introduced. It was shown that the generator could easily recover the specific private data of users, particularly client identity.

An efficient and robust protocol for high-dimensional data secure aggregation was proposed in [18] that can be used in FL. Using this protocol, a server was able to compute the sum of large user-held data vectors from mobile devices to aggregate user-provided model updates for a DNN model without distinguishing individual user's contributions. In addition, the effectiveness and efficiency of the proposed protocol were verified.

A GAN-based privacy-preserving method was proposed in [155] to obfuscate users' sensitive information. The proposed method employed a generator to produce an optimal obfuscation method for data protection. At the same time, a classifier was used to deobfuscate the data. These two nets continued to play against each other until they achieved an equilibrium. This process can raise the level of protection. By investigating location privacy protection on the Gowalla dataset and synthetic data, it was shown that the proposed approach could achieve privacy protection and deal with the Bayes error.

To alleviate the threat of black-box inference attacks against ML models, a mechanism to train models with membership privacy was introduced in [137]. By formulating a min-max game, an adversarial training algorithm was designed to minimize the prediction loss of the model and the maximum gain of the inference attacks. The effectiveness of the min-max strategy on defending membership inference attacks was verified without significantly downgrading the model's prediction accuracy.

To defend against the attribute inference attacks, a countermeasure named AttriGuard was proposed in [90]. AttriGuard works in two phases. In Phase I, the minimum noise was found by adapting existing evasion attacks in adversarial ML. This noise protects users' attribute values by adding itself to the user's public data. In Phase II, the proposed method sampled one attribute value according to a certain probability distribution and added the corresponding noise found in Phase I to the user's public data.

Blockchain techniques offer new security and privacy protection options to the distributed network. It can enhance security and privacy protection by providing authentication, access control, and confidentiality [58]. The authors in [5] developed a blockchain-based method to protect the user's security in the SS network. A Multiple Operators SS (MOSS) smart contract framework was proposed to allow users to share their spectrum decentralized and secure.

Moreover, a combination of blockchain and a DRL framework named DeepCoin has been proposed in [57] to preserve the energy system to defend against Byzantine attacks. By using short signatures and functions to generate the blocks, it can prevent smart grid attacks. Furthermore, the DRL in their framework can detect network attacks and fraudulent transactions for the blockchain-based energy network by using recurrent neural networks. The performance of their proposed method has been verified.

11.3 Summary

In this chapter, a comprehensive investigation of state-of-the- art ML-based SS solutions was presented. It has been noted that the dynamic access and sharing paradigms of SS networks may open the system to many security concerns. Two typical spectrum sensing attacks were discussed, i.e. PUE and SSDF. Two common attacks, i.e. jamming and eavesdropping, during wireless access and transmission in the context of the SS network were also addressed. Furthermore, the coexistence of different types of users and the application of ML all require massive information exchanges, generating tremendous concerns about privacy.

12

Open Issues and Future Directions for 5G and Beyond Wireless Networks

Wireless communication is a fast-growing field that attracts much research attention, from both industry and academia. As we progress to the next decade, the need for wireless connectivity will continue to grow, and the big challenge arises when the expected wireless system support new applications, such as metaverse, vehicle communication, and at space-air-ground domain. The demand–supply relationship in wireless communication will continue to push technologies forward. In this chapter, we briefly summarize active research for the next-generation wireless systems.

12.1 Joint Communication and Sensing

Future wireless system not only desires for highly reliable and faster connectivity, but also the ability to sense surrounding dynamics with ubiquitous wireless signals [133]. Such sensing capabilities include range and velocity estimation, object detection, collision avoidance, and localization. In the past, communication and sensing have been designed separately, where either communication or sensing is the main target design, the other as the by-product. However, integrating both functionalities within one system has clear advantages in providing better power and spectral efficiency, as well as reducing hardware and signaling cost through efficient resource coordination [56, 117].

Joint communication and sensing has attracted extensive attention from the perspective of jointly considering and unifying two operations, especially in emerging applications like vehicle-to-everything (V2X), where simultaneous information exchange and radar-like parameter estimation are critical. Yet, achieving integration gain from Joint communication and sensing system faces numerous challenges. Among them, Joint communication and sensing physical layer design such as waveform optimization, collaborative resource allocation, and beamforming are of paramount importance.

12.2 Space-Air-Ground Communication

One limitation for 5G wireless system is that it provides network access mainly for terrestrial (ground) communication. Emerging applications in the space, such as satellite, and

5G and Beyond Wireless Communication Networks, First Edition. Haijian Sun, Rose Qingyang Hu, and Yi Qian.
© 2024 John Wiley & Sons Ltd. Published 2024 by John Wiley & Sons Ltd.

air communication for UAVs have different needs. Besides, their design philosophy is very much different. For example, UAV communication expects LoS scenarios (signal propagation is in favor condition) but power and trajectory optimization are more important design factors. Recently, the space-air-ground integrated network (SAGIN) has been proposed. SAGIN provides seamless coverage for larger areas, including sea, space, ground, and air.

SAGIN has to consider various factors from each segment, current design focuses on protocol optimization, resource allocation, performance analysis, mobility management, and inter-segment operation [112]. Furthermore, the network design and system integration in SAGIN are of great significance.

12.3 Semantic Communication

Communication should be used not only for exchanging data bits, but also for semantic exchange. In fact, many scenarios involve semantic information. For example, transmitting natural languages. Current solution needs to convert language into bits via upper layer operations, then simply send those bits through medium. In [150], the difference between bit and semantic transmission is shown in Figure 1 therein. Essentially, semantic communication will utilize advanced ML techniques to perform semantic encoding at transmitter and correspondingly semantic decoding at the receiver end, avoiding the source encoding/decoding directly. The challenge is that current system will undergo significant modifications, for example, the need for new metric for semantic entropy, semantic channel, and noise factor. Nevertheless, semantic communication is regarded as an important component in 5G beyond.

12.4 Data-Driven Communication System Design

The exponential growth of data traffic and recent advancement on ML fuel the data-driven communication system design. In particular, collected data can help innovative designs on modulation, coding, scheduling, architecture, resource management, and even end-to-end [217]. ML is one of the most powerful tools and can effectively explore massive data and make accurate predictions and plannings. For example, future communication design may not start from problem formulation and then be solved with traditional convex (or non-convex) optimization. Rather, the model design can learn from data and reach something we have yet seen. Another big challenge that data-driven approach can address is the network scalability, which is of critical importance in massive IoT era.

Appendix A

Proof of Theorem 5.1

To prove the Theorem, we first consider the KKT conditions of $\mathbf{P}_3$. Specifically, with some simple algebraic manipulation, (5.11) can be rewritten as

$$
\begin{bmatrix} \alpha_{i,k}\mathbf{I} & \mathbf{0} \\ \mathbf{0} & t_{i,k} \end{bmatrix} + \begin{bmatrix} \mathbf{I} \\ \hat{\mathbf{h}}_i^{\dagger} \end{bmatrix} \mathbf{C}_k \begin{bmatrix} \mathbf{I} & \hat{\mathbf{h}}_i \end{bmatrix} + \begin{bmatrix} -\gamma_{k,\min}\sum_{j=1}^{k-1}\mathbf{W}_j & \mathbf{0} \\ \mathbf{0} & 0 \end{bmatrix}
$$

$$
\succeq \mathbf{0}, \ \forall k \in \mathcal{K}, \ i = \{k, k+1, \dots, K\}, \tag{A.1}
$$

where $t_{i,k} = -\alpha_{i,k}\varphi_k^2 - \gamma_{k,\min}\left(\sigma_{k,S}^2 + \frac{\sigma_D^2}{(1-\rho)}\right)$.

Similarly, (5.13) and 5.15 can be rewritten as

$$
\begin{bmatrix} \beta_n\mathbf{I} & \mathbf{0} \\ \mathbf{0} & -\beta_n\psi_n^2 + P_{n,p} \end{bmatrix} - \begin{bmatrix} \mathbf{I} \\ \hat{\mathbf{g}}_n^{\dagger} \end{bmatrix} \mathbf{\Sigma} \begin{bmatrix} \mathbf{I} & \hat{\mathbf{g}}_n \end{bmatrix} \succeq \mathbf{0}, \ \forall n \in \mathcal{N}, \tag{A.2}
$$

and

$$
\begin{bmatrix} \theta_k\mathbf{I} & \mathbf{0} \\ \mathbf{0} & m_k \end{bmatrix} + \begin{bmatrix} \mathbf{I} \\ \hat{\mathbf{h}}_k^{\dagger} \end{bmatrix} \mathbf{\Sigma} \begin{bmatrix} \mathbf{I} & \hat{\mathbf{h}}_k \end{bmatrix} \succeq \mathbf{0}, \ \forall k \in \mathcal{K}, \tag{A.3}
$$

respectively, where $m_k = -\theta_k\varphi_k^2 + \sigma_{k,S}^2 - \frac{\tau_k}{\rho}$.

For notational simplicity, we let $\mathbf{X}_i = \begin{bmatrix} \mathbf{I} & \hat{\mathbf{h}}_i \end{bmatrix}$ and $\mathbf{Y}_n = \begin{bmatrix} \mathbf{I} & \hat{\mathbf{g}}_n \end{bmatrix}$. Also, denote $\mathbf{A}_{i,k} \in \mathbb{C}_+^{(M+1) \times (M+1)}$, $\mathbf{B}_k \in \mathbb{C}_+^{(M+1) \times (M+1)}$, $\mathbf{D}_n \in \mathbb{C}_+^{(M+1) \times (M+1)}$, $z \in \mathbb{R}_+$, and $\mathbf{E}_k \in \mathbb{C}_+^{(M) \times (M)}$ as the KKT multiplier. Then the Lagrange dual function $\mathcal{L}$ can be expressed as

$$
\mathcal{L}(\mathbf{W}_k, \mathbf{V}, \mathbf{A}_{i,k}, \mathbf{B}_k, \mathbf{D}_n, z, \kappa) = \mathrm{Tr}(\mathbf{\Sigma}) - \sum_{i,k} \mathrm{Tr}(\mathbf{A}_{i,k}\mathbf{X}_i^{\dagger}\mathbf{C}_k\mathbf{X}_i)
$$

$$
- \sum_{i,k} \mathrm{Tr}(\mathbf{A}_{i,k}\mathbf{M}_k) + \sum_n \mathrm{Tr}(\mathbf{D}_n\mathbf{Y}_n^{\dagger}\mathbf{\Sigma}\mathbf{Y}_n) - \sum_k \mathrm{Tr}\left(\mathbf{B}_k\mathbf{X}_k^{\dagger}\mathbf{\Sigma}\mathbf{X}_k\right) +
$$

$$
z\left(\mathrm{Tr}(\mathbf{\Sigma}) - P_B\right) - \sum_k \mathrm{Tr}(\mathbf{E}_k\mathbf{W}_k) + \kappa, \tag{A.4}
$$

where $\mathbf{M}_k = \begin{bmatrix} -\gamma_{k,\min}\sum_{j=1}^{k-1}\mathbf{W}_j & \mathbf{0} \\ \mathbf{0} & 0 \end{bmatrix}$ and κ are the terms irrelevant of $\mathbf{W}_k$. Taking the partial derivative of the dual function regarding $\mathbf{W}_k$, we have

$$\frac{\partial \mathcal{L}}{\partial \mathbf{W}_k} = \mathbf{I} - \sum_i \mathbf{X}_i \mathbf{A}_{i,k} \mathbf{X}_i^\dagger + \gamma_{k,\min} \sum_i \sum_{j=1}^{k-1} \mathbf{X}_i \mathbf{A}_{i,j} \mathbf{X}_i^\dagger$$

$$+ \sum_i \gamma_{k,\min} \sum_{j=k+1}^{K} \mathbf{A}_{i,j} + \sum_n \mathbf{Y}_n \mathbf{D}_n \mathbf{Y}_n^\dagger - \sum_k \mathbf{X}_k \mathbf{B}_k \mathbf{X}_k^\dagger + z\mathbf{I} - \mathbf{E}_k = \mathbf{0}. \tag{A.5}$$

In addition, the dual problem needs to satisfy the completeness slackness

$$\left(\begin{bmatrix} \alpha_{i,k}\mathbf{I} & \mathbf{0} \\ \mathbf{0} & t_{i,k} \end{bmatrix} + \mathbf{X}_i^\dagger \mathbf{C}_k \mathbf{X}_i + \mathbf{M}_k \right) \mathbf{A}_{i,k} = \mathbf{0}, \tag{A.6a}$$

$$\mathbf{E}_k \mathbf{W}_k = \mathbf{0}, \forall k \in \mathcal{K}, i = \{k+1, \dots, K\}, \forall n \in \mathcal{N}. \tag{A.6b}$$

Right multiplying $\mathbf{W}_k$ with (A.5), and substituting (A.6b), we can get

$$\left(\sum_i \mathbf{X}_i \mathbf{A}_{i,k} \mathbf{X}_i^\dagger + \sum_k \mathbf{X}_k \mathbf{B}_k \mathbf{X}_k^\dagger \right) \mathbf{W}_k = \left[(1+z)\mathbf{I} + \gamma_{k,\min} \sum_i \sum_{j=k+1}^{K} \mathbf{A}_{i,j} \right.$$

$$\left. + \gamma_{k,\min} \sum_i \sum_{j=1}^{k-1} \mathbf{X}_i \mathbf{A}_{i,j} \mathbf{X}_i^\dagger + \sum_n \mathbf{Y}_n \mathbf{D}_n \mathbf{Y}_n^\dagger \right] \mathbf{W}_k. \tag{A.7}$$

Since all the KKT multipliers are positive numbers or positive semidefinite matrix, $\left\{ (1+z)\mathbf{I} + \gamma_{k,\min} \sum_i \sum_{j=1}^{k-1} \mathbf{X}_i \mathbf{A}_{i,j} \mathbf{X}_i^\dagger + \gamma_{k,\min} \sum_i \sum_{j=k+1}^{K} \mathbf{A}_{i,j} + \sum_n \mathbf{Y}_n \mathbf{D}_n \mathbf{Y}_n^\dagger \right\} \geq \mathbf{0}$. Thus it is non-singular. Left multiplying a non-singular matrix with $\mathbf{W}_k$ does not change the rank of $\mathbf{W}_k$. Therefore, we have

$$\text{Rank}(\mathbf{W}_k) = \text{Rank}\left(\left(\sum_i \mathbf{X}_i \mathbf{A}_{i,k} \mathbf{X}_i^\dagger + \sum_k \mathbf{X}_k \mathbf{B}_k \mathbf{X}_k^\dagger \right) \mathbf{W}_k \right)$$

$$= \min \left\{ \text{Rank}\left(\sum_i \mathbf{X}_i \mathbf{A}_{i,k} \mathbf{X}_i^\dagger + \sum_k \mathbf{X}_k \mathbf{B}_k \mathbf{X}_k^\dagger \right), \text{Rank}(\mathbf{W}_k) \right\}. \tag{A.8}$$

Next, we show the rank of $(\sum_i \mathbf{X}_i \mathbf{A}_{i,k} \mathbf{X}_i^\dagger)$ is less than or equal to 2. By summing (A.6a) in terms of index i, then left-multiplying $\begin{bmatrix} \mathbf{I}_M & \mathbf{0} \end{bmatrix}$ and right-multiplying $\mathbf{X}_i^\dagger$, we have

$$\sum_i \alpha_{i,k} \mathbf{X}_i \mathbf{A}_{i,k} \mathbf{X}_i^\dagger - \sum_i \alpha_{i,k} \begin{bmatrix} \mathbf{0}_M & \mathbf{h}_i \end{bmatrix} \mathbf{A}_{i,k} \mathbf{X}_i^\dagger$$

$$+ \sum_i \mathbf{C}_k \mathbf{X}_i \mathbf{A}_{i,k} \mathbf{X}_i^\dagger + \sum_i \left(-\gamma_{k,\min} \sum_{j=1}^{k-1} \mathbf{W}_j \right) \mathbf{X}_i \mathbf{A}_{i,k} \mathbf{X}_i^\dagger$$

$$- \sum_i \left(-\gamma_{k,\min} \sum_{j=1}^{k-1} \mathbf{W}_j \right) \begin{bmatrix} \mathbf{0}_M & \mathbf{h}_i \end{bmatrix} \mathbf{A}_{i,k} \mathbf{X}_i^\dagger = \mathbf{0}. \tag{A.9}$$

After a simple transformation, we have

$$\sum_i \left(\alpha_{i,k}\mathbf{I} + \mathbf{C}_k - \gamma_{k,\min} \sum_{j=1}^{k-1} \mathbf{W}_j \right) \mathbf{X}_i \mathbf{A}_{i,k} \mathbf{X}_i^\dagger$$

$$= \sum_i \left(\alpha_{i,k}\mathbf{I} - \gamma_{k,\min} \sum_{j=1}^{k-1} \mathbf{W}_j \right) \begin{bmatrix} \mathbf{0}_M & \mathbf{h}_i \end{bmatrix} \mathbf{A}_{i,k} \mathbf{X}_i^\dagger. \tag{A.10}$$

From the fact that (5.11) is a positive semidefinite matrix, $\left(\alpha_{i,k} \mathbf{I} + \mathbf{C}_k - \gamma_{k,\min} \sum_{j=1}^{k-1} \mathbf{W}_j \right)$ would be a non-singular matrix, thus the rank of the left term of the above equation is the same as $\sum_i \mathbf{X}_i \mathbf{A}_{i,k} \mathbf{X}_i^\dagger$. Also, it is easy to verify that the right term has a rank 1.

Similarly, we can prove that $\text{Rank}(\sum_n \mathbf{Y}_n \mathbf{D}_n \mathbf{Y}_n^\dagger) = 1$. Therefore, the following equation holds.

$$\text{Rank}\left(\sum_i \mathbf{X}_i \mathbf{A}_{i,k} \mathbf{X}_i^\dagger + \sum_k \mathbf{X}_k \mathbf{B}_k \mathbf{X}_k^\dagger \right) \leq \text{Rank}\left(\sum_i \mathbf{X}_i \mathbf{A}_{i,k} \mathbf{X}_i^\dagger \right)$$

$$+\text{Rank}\left(\sum_n \mathbf{Y}_n \mathbf{D}_n \mathbf{Y}_n^\dagger \right) = 2, \tag{A.11}$$

which proves the theorem.

Bibliography

1 M. E. Ahmed, J. B. Song, and Z. Han. Mitigating malicious attacks using Bayesian nonparametric clustering in collaborative cognitive radio networks. In *2014 IEEE Global Communications Conference*, pages 999–1004, 2014.

2 I. F. Akyildiz, W.-Y. Lee, M. C. Vuran, and S. Mohanty. Next generation/dynamic spectrum access/cognitive radio wireless networks: A survey. *Computer Networks*, 50(13):2127–2159, 2006.

3 M. M. Al-Wani, A. Sali, B. M. Ali, A. A. Salah, K. Navaie, C. Y. Leow, N. K. Noordin, and S. J. Hashim. On short term fairness and throughput of user clustering for downlink non-orthogonal multiple access system. In *2019 IEEE 89th Vehicular Technology Conference (VTC2019-Spring)*, pages 1–6, 2019. doi: 10.1109/VTCSpring.2019.8746330.

4 A. Albehadili, A. Ali, F. Jahan, A. Y. Javaid, J. Oluochy, and V. Devabhaktuniz. Machine learning-based primary user emulation attack detection in cognitive radio networks using pattern described link-signature (PDLS). In *2019 Wireless Telecommunications Symposium (WTS)*, pages 1–7, 2019.

5 H. Alhosani, M. H. ur Rehman, K. Salah, C. Lima, and D. Svetinovic. Blockchain-based solution for multiple operator spectrum sharing (MOSS) in 5G networks. In *2020 IEEE Globecom Workshops (GC Wkshps*, pages 1–6, 2020. doi: 10.1109/GCWkshps50303.2020.9367561.

6 M. M. Amiri and D. Gündüz. Machine learning at the wireless edge: Distributed stochastic gradient descent over-the-air. *IEEE Transactions on Signal Processing*, 68:2155–2169, 2020.

7 S. Amuru and R. M. Buehrer. Optimal jamming using delayed learning. In *2014 IEEE Military Communications Conference*, pages 1528–1533, 2014.

8 J. G. Andrews and T. H. Meng. Optimum power control for successive interference cancellation with imperfect channel estimation. *IEEE Transactions on Wireless Communications*, 2(2):375–383, 2003.

9 J. G. Andrews and T. H.-Y. Meng. Performance of multicarrier CDMA with successive interference cancellation in a multipath fading channel. *IEEE Transactions on Communications*, 52(5):811–822, 2004.

5G and Beyond Wireless Communication Networks, First Edition. Haijian Sun, Rose Qingyang Hu, and Yi Qian.
© 2024 John Wiley & Sons Ltd. Published 2024 by John Wiley & Sons Ltd.

10 J. G. Andrews, S. Buzzi, W. Choi, S. V. Hanly, A. Lozano, A. C. K. Soong, and J. C. Zhang. What will 5G be? *IEEE Journal on Selected Areas in Communications*, 32(6):1065–1082, 2014.

11 M. A. Aref, S. K. Jayaweera, and S. Machuzak. Multi-agent reinforcement learning based cognitive anti-jamming. In *2017 IEEE Wireless Communications and Networking Conference (WCNC)*, pages 1–6, 2017.

12 S. Arun and G. Umamaheswari. An adaptive learning-based attack detection technique for mitigating primary user emulation in cognitive radio networks. *Circuits, Systems, and Signal Processing*, 39(2):1071–1088, 2020. ISSN 1531-5878.

13 I. Bechar. A Bernstein-type inequality for stochastic processes of quadratic forms of Gaussian variables. *arXiv preprint arXiv:0909.3595*, 2009.

14 S. Bhattarai, P. R. Vaka, and J. Park. Thwarting location inference attacks in database-driven spectrum sharing. *IEEE Transactions on Cognitive Communications and Networking*, 4(2):314–327, 2018.

15 S. Bi and Y. J. Zhang. Computation rate maximization for wireless powered mobile-edge computing with binary computation offloading. *IEEE Transactions on Wireless Communications*, 17(6):4177–4190, 2018.

16 V. Bioglio, C. Condo, and I. Land. Design of polar codes in 5G new radio. *IEEE Communications Surveys & Tutorials*, 23(1):29–40, 2020.

17 M. Bkassiny, Y. Li, and S. K. Jayaweera. A survey on machine-learning techniques in cognitive radios. *IEEE Communications Surveys & Tutorials*, 15(3):1136–1159, 2013. doi: 10.1109/SURV.2012.100412.00017.

18 K. Bonawitz, V. Ivanov, B. Kreuter, A. Marcedone, H. B. McMahan, S. Patel, D. Ramage, A. Segal, and K. Seth. Practical secure aggregation for privacy-preserving machine learning. In *Proceedings of the 2017 ACM SIGSAC Conference on Computer and Communications Security*, pages 1175–1191, 2017.

19 E. Boshkovska, D. W. K. Ng, N. Zlatanov, and R. Schober. Practical non-linear energy harvesting model and resource allocation for SWIPT systems. *IEEE Communications Letters*, 19(12):2082–2085, 2015. doi: 10.1109/LCOMM.2015.2478460.

20 E. Boshkovska, A. Koelpin, D. W. K. Ng, N. Zlatanov, and R. Schober. Robust beamforming for SWIPT systems with non-linear energy harvesting model. In *2016 IEEE 17th International Workshop on Signal Processing Advances in Wireless Communications (SPAWC)*, pages 1–5, 2016. doi: 10.1109/SPAWC.2016.7536860.

21 E. Boshkovska, R. Morsi, D. W. K. Ng, and R. Schober. Power allocation and scheduling for SWIPT systems with non-linear energy harvesting model. In *2016 IEEE International Conference on Communications (ICC)*, pages 1–6, 2016. doi: 10.1109/ICC.2016.7511403.

22 E. Boshkovska, D. W. K. Ng, N. Zlatanov, A. Koelpin, and R. Schober. Robust resource allocation for MIMO wireless powered communication networks based on a non-linear EH model. *IEEE Transactions on Communications*, 65(5):1984–1999, 2017. doi: 10.1109/TCOMM.2017.2664860.

23 E. Boshkovska, N. Zlatanov, L. Dai, D. W. K. Ng, and R. Schober. Secure SWIPT networks based on a non-linear energy harvesting model. In *2017 IEEE Wireless*

Communications and Networking Conference Workshops (WCNCW), pages 1–6, 2017. doi: 10.1109/WCNCW.2017.7919062.

24 E. Boshkovska, D. W. K. Ng, L. Dai, and R. Schober. Power-efficient and secure WPCNs with hardware impairments and non-linear EH circuit. *IEEE Transactions on Communications*, 66(6):2642–2657, 2018. doi: 10.1109/TCOMM.2017.2783628.

25 S. Boyd and L. Vandenberghe. *Convex Optimization*. Number pt. 1 in Berichte über verteilte messyteme. Cambridge University Press, 2004. ISBN 9780521833783. URL https://books.google.com/books?id=mYm0bLd3fcoC.

26 F. Cai, Y. Gao, L. Cheng, L. Sang, and D. Yang. Spectrum sharing for LTE and WiFi coexistence using decision tree and game theory. In *2016 IEEE Wireless Communications and Networking Conference*, pages 1–6, 2016.

27 Y. Cai, K. Shi, F. Song, Y. Xu, X. Wang, and H. Luan. Jamming pattern recognition using spectrum waterfall: A deep learning method. In *2019 IEEE 5th International Conference on Computer and Communications (ICCC)*, pages 2113–2117, 2019.

28 U. Challita, L. Dong, and W. Saad. Proactive resource management for LTE in unlicensed spectrum: A deep learning perspective. *IEEE Transactions on Wireless Communications*, 17(7):4674–4689, 2018.

29 X. Chen and J. Huang. Database-assisted distributed spectrum sharing. *IEEE Journal on Selected Areas in Communications*, 31(11):2349–2361, 2013.

30 X. Chen, D. W. K. Ng, W. H. Gerstacker, and H.-H. Chen. A survey on multiple-antenna techniques for physical layer security. *IEEE Communications Surveys & Tutorials*, 19(2):1027–1053, 2017. doi: 10.1109/COMST.2016.2633387.

31 M. Chen, W. Saad, and C. Yin. Echo state networks for self-organizing resource allocation in LTE-U with uplink-downlink decoupling. *IEEE Transactions on Wireless Communications*, 16(1):3–16, 2017.

32 M. Chen, W. Saad, and C. Yin. Liquid state machine learning for resource allocation in a network of cache-enabled LTE-U UAVs. In *GLOBECOM 2017 - 2017 IEEE Global Communications Conference*, pages 1–6, 2017.

33 M. Chen, W. Saad, and C. Yin. Liquid state machine learning for resource and cache management in LTE-U unmanned aerial vehicle (UAV) networks. *IEEE Transactions on Wireless Communications*, 18(3):1504–1517, 2019.

34 Cisco. Cisco visual networking index: Global mobile data traffic forecast update, 2016–2021 white paper. 2017.

35 T. C. Clancy, A. Khawar, and T. R. Newman. Robust signal classification using unsupervised learning. *IEEE Transactions on Wireless Communications*, 10(4):1289–1299, 2011.

36 M. Clark and K. Psounis. Optimizing primary user privacy in spectrum sharing systems. *IEEE/ACM Transactions on Networking*, 28(2):533–546, 2020. doi: 10.1109/TNET.2020.2967776.

37 M. Dabaghchian, A. Alipour-Fanid, K. Zeng, and Q. Wang. Online learning-based optimal primary user emulation attacks in cognitive radio networks. In *2016 IEEE Conference on Communications and Network Security (CNS)*, pages 100–108, 2016.

38 M. Dabaghchian, A. Alipour-Fanid, K. Zeng, Q. Wang, and P. Auer. Online learning with randomized feedback graphs for optimal PUE attacks in cognitive radio networks. *IEEE/ACM Transactions on Networking*, 26(5):2268–2281, 2018.

39 K. Davaslioglu and Y. E. Sagduyu. Generative adversarial learning for spectrum sensing. In *2018 IEEE International Conference on Communications (ICC)*, pages 1–6, 2018.

40 N. Devroye, P. Mitran, and V. Tarokh. Cognitive decompiosition of wirless networks. In *2006 1st International Conference on Cognitive Radio Oriented Wireless Networks and Communications*, pages 1–5, 2006.

41 P. D. Diamantoulakis, K. N. Pappi, Z. Ding, and G. K. Karagiannidis. Wireless-powered communications with non-orthogonal multiple access. *IEEE Transactions on Wireless Communications*, 15(12):8422–8436, 2016.

42 P. D. Diamantoulakis, K. N. Pappi, G. K. Karagiannidis, H. Xing, and A. Nallanathan. Joint downlink/uplink design for wireless powered networks with interference. *IEEE Access*, 5:1534–1547, 2017.

43 A. Dimas, M. A. Clark, B. Li, K. Psounis, and A. P. Petropulu. On radar privacy in shared spectrum scenarios. In *ICASSP 2019 - 2019 IEEE International Conference on Acoustics, Speech and Signal Processing (ICASSP)*, pages 7790–7794, 2019. doi: 10.1109/ICASSP.2019.8682745.

44 Z. Ding, Z. Yang, P. Fan, and H. V. Poor. On the performance of non-orthogonal multiple access in 5G systems with randomly deployed users. *IEEE Signal Processing Letters*, 21(12):1501–1505, 2014.

45 Z. Ding, P. Fan, and H. V. Poor. Impact of user pairing on 5G nonorthogonal multiple-access downlink transmissions. *IEEE Transactions on Vehicular Technology*, 65(8):6010–6023, 2015.

46 Z. Ding, H. Dai, and H. V. Poor. Relay selection for cooperative NOMA. *IEEE Wireless Communications Letters*, 5(4):416–419, 2016. doi: 10.1109/LWC.2016 .2574709.

47 Z. Ding, P. Fan, and H. V. Poor. Random beamforming in millimeter-wave NOMA networks. *IEEE Access*, 5:7667–7681, 2017.

48 V. Q. Do and I. Koo. Learning frameworks for cooperative spectrum sensing and energy-efficient data protection in cognitive radio networks. *Applied Sciences*, 8(5):722, 2018.

49 N. T. Do, D. B. Da Costa, T. Q. Duong, and B. An. A BNBF user selection scheme for NOMA-based cooperative relaying systems with SWIPT. *IEEE Communications Letters*, 21(3):664–667, 2017. doi: 10.1109/LCOMM.2016.2631606.

50 Q. Dong, Y. Chen, X. Li, and K. Zeng. Explore recurrent neural network for PUE attack detection in practical CRN models. In *2018 IEEE International Smart Cities Conference (ISC2)*, pages 1–9, 2018.

51 K. Doppler, M. Rinne, C. Wijting, C. B. Ribeiro, and K. Hugl. Device-to-device communication as an underlay to LTE-advanced networks. *IEEE Communications Magazine*, 47(12):42–49, 2009. doi: 10.1109/MCOM.2009.5350367.

52 S. M. Elghamrawy. Security in cognitive radio network: Defense against primary user emulation attacks using genetic artificial bee colony (GABC) algorithm.

Future Generation Computer Systems, 109:479–487, 2020. ISSN 0167-739X. doi: https://doi.org/10.1016/j.future.2018.08.022. URL http://www.sciencedirect.com/science/article/pii/S0167739X17321246.

53 M. Fahimi and A. Ghasemi. A distributed learning automata scheme for spectrum management in self-organized cognitive radio network. *IEEE Transactions on Mobile Computing*, 16(6):1490–1501, 2017.

54 B. Fang, Z. Qian, W. Zhong, and W. Shao. AN-aided secrecy precoding for SWIPT in cognitive MIMO broadcast channels. *IEEE Communications Letters*, 19(9):1632–1635, 2015.

55 F. Farmani, M. Abbasi-Jannatabad, and R. Berangi. Detection of SSDF attack using SVDD algorithm in cognitive radio networks. In *2011 Third International Conference on Computational Intelligence, Communication Systems and Networks*, pages 201–204, 2011.

56 Z. Feng, Z. Fang, Z. Wei, X. Chen, Z. Quan, and D. Ji. Joint radar and communication: A survey. *China Communications*, 17(1):1–27, 2020.

57 M. A. Ferrag and L. Maglaras. DeepCoin: A novel deep learning and blockchain-based energy exchange framework for smart grids. *IEEE Transactions on Engineering Management*, 67(4):1285–1297, 2020. doi: 10.1109/TEM.2019.2922936.

58 M. A. Ferrag and L. Shu. The performance evaluation of blockchain-based security and privacy systems for the Internet of Things: A tutorial. *IEEE Internet of Things Journal*, 8(24):17236–17260, 2021. doi: 10.1109/JIOT.2021.3078072.

59 M. P. C. Fossorier. Quasicyclic low-density parity-check codes from circulant permutation matrices. *IEEE Transactions on Information Theory*, 50(8):1788–1793, 2004. doi: 10.1109/TIT.2004.831841.

60 H. M. Furqan, M. A. Aygül, M. Nazzal, and H. Arslan. Primary user emulation and jamming attack detection in cognitive radio via sparse coding. *EURASIP Journal on Wireless Communications and Networking*, 2020(1):141, 2020. ISSN 1687-1499.

61 Z. Gao, H. Zhu, S. Li, S. Du, and X. Li. Security and privacy of collaborative spectrum sensing in cognitive radio networks. *IEEE Wireless Communications*, 19(6):106–112, 2012. doi: 10.1109/MWC.2012.6393525.

62 J. Geiping, H. Bauermeister, H. Dröge, and M. Moeller. Inverting gradients–how easy is it to break privacy in federated learning? *arXiv preprint arXiv:2003.14053*, 2020.

63 A. Goldsmith, S. A. Jafar, N. Jindal, and S. Vishwanath. Capacity limits of MIMO channels. *IEEE Journal on Selected Areas in Communications*, 21(5):684–702, 2003.

64 A. Goldsmith, S. A. Jafar, I. Maric, and S. Srinivasa. Breaking spectrum gridlock with cognitive radios: An information theoretic perspective. *Proceedings of the IEEE*, 97(5):894–914, 2009. doi: 10.1109/JPROC.2009.2015717.

65 I. S. Gradshteyn and I. M. Ryzhik. *Table of Integrals, Series, and Products*. Academic Press, 2014.

66 M. Grissa, A. A. Yavuz, and B. Hamdaoui. Location privacy preservation in database-driven wireless cognitive networks through encrypted probabilistic data structures. *IEEE Transactions on Cognitive Communications and Networking*, 3(2):255–266, 2017.

67 H. Guo, Q. Zhang, S. Xiao, and Y. Liang. Exploiting multiple antennas for cognitive ambient backscatter communication. *IEEE Internet of Things Journal*, 6(1):765–775, 2019.

68 A. K. Gupta and D. K. Nagar. *Matrix Variate Distributions*. Chapman and Hall/CRC, 2018.

69 Y. Gwon, S. Dastangoo, C. Fossa, and H. T. Kung. Competing mobile network game: Embracing antijamming and jamming strategies with reinforcement learning. In *2013 IEEE Conference on Communications and Network Security (CNS)*, pages 28–36, 2013.

70 A. Hassan, R. Hamza, H. Yan, and P. Li. An efficient outsourced privacy preserving machine learning scheme with public verifiability. *IEEE Access*, 7:146322–146330, 2019. doi: 10.1109/ACCESS.2019.2946202.

71 T. Hayashida, R. Okumura, K. Mizutani, and H. Harada. Possibility of dynamic spectrum sharing system by VHF-band radio sensor and machine learning. In *2019 IEEE International Symposium on Dynamic Spectrum Access Networks (DySPAN)*, pages 1–6, 2019.

72 D. T. Hoang, D. Niyato, P. Wang, D. I. Kim, and Z. Han. Ambient backscatter: A new approach to improve network performance for RF-powered cognitive radio networks. *IEEE Transactions on Communications*, 65(9):3659–3674, 2017.

73 M. A. Hossain, R. M. Noor, K. A. Yau, S. R. Azzuhri, M. R. Z'aba, and I. Ahmedy. Comprehensive survey of machine learning approaches in cognitive radio-based vehicular Ad Hoc networks. *IEEE Access*, 8:78054–78108, 2020.

74 M. A. Hossen and S. Yoo. Q-learning based multi-objective clustering algorithm for cognitive radio Ad Hoc networks. *IEEE Access*, 7:181959–181971, 2019.

75 M. Höyhtyä, A. Mämmelä, X. Chen, A. Hulkkonen, J. Janhunen, J. Dunat, and J. Gardey. Database-assisted spectrum sharing in satellite communications: A survey. *IEEE Access*, 5:25322–25341, 2017.

76 R. Q. Hu and Y. Qian. *Heterogeneous Cellular Networks*. John Wiley & Sons, 2013.

77 R. Q. Hu and Y. Qian. *Resource Management for Heterogeneous Networks in LTE Systems*. Springer, 2014.

78 R. Q. Hu and Y. Qian. An energy efficient and spectrum efficient wireless heterogeneous network framework for 5G systems. *IEEE Communications Magazine*, 52(5):94–101, 2014.

79 Z. Hu, N. Wei, and Z. Zhang. Optimal resource allocation for harvested energy maximization in wideband cognitive radio network with SWIPT. *IEEE Access*, 5:23383–23394, 2017.

80 Y. Hu, P. Wang, Z. Lin, M. Ding, and Y. Liang. Machine learning based signal detection for ambient backscatter communications. In *ICC 2019 - 2019 IEEE International Conference on Communications (ICC)*, pages 1–6, 2019.

81 X. Huang, T. Han, and N. Ansari. On green-energy-powered cognitive radio networks. *IEEE Communications Surveys & Tutorials*, 17(2):827–842, 2015.

82 C. Huang, G. Chen, Y. Gong, and P. Xu. Deep reinforcement learning based relay selection in delay-constrained secure buffer-aided CRNs. In *GLOBECOM 2020 - 2020 IEEE Global Communications Conference*, pages 1–6, 2020. doi: 10.1109/GLOBECOM42002.2020.9322098.

83 Y. Huo, Y. Wang, W. Lin, and R. Sun. Three-layer Bayesian model based spectrum sensing to detect malicious attacks in cognitive radio networks. In *2015 IEEE International Conference on Communication Workshop (ICCW)*, pages 1640–1645, 2015.

84 N. V. Huynh, D. T. Hoang, D. N. Nguyen, E. Dutkiewicz, D. Niyato, and P. Wang. Reinforcement learning approach for RF-powered cognitive radio network with ambient backscatter. In *2018 IEEE Global Communications Conference (GLOBECOM)*, pages 1–6, 2018.

85 N. V. Huynh, D. N. Nguyen, D. T. Hoang, E. Dutkiewicz, M. Mueck, and S. Srikanteswara. Defeating jamming attacks with ambient backscatter communications. In *2020 International Conference on Computing, Networking and Communications (ICNC)*, pages 405–409, 2020.

86 M. A. Inamdar and H. V. Kumaraswamy. Accurate primary user emulation attack (PUEA) detection in cognitive radio network using KNN and ANN classifier. In *2020 4th International Conference on Trends in Electronics and Informatics (ICOEI)(48184)*, pages 490–495, 2020.

87 F. Jameel, W. U. Khan, S. T. Shah, and T. Ristaniemi. Towards intelligent IoT networks: Reinforcement learning for reliable backscatter communications. In *2019 IEEE Globecom Workshops (GC Wkshps)*, pages 1–6, 2019.

88 F. Jameel, M. A. Jamshed, Z. Chang, R. Jäntti, and H. Pervaiz. Low latency ambient backscatter communications with deep Q-learning for beyond 5G applications. In *2020 IEEE 91st Vehicular Technology Conference (VTC2020-Spring)*, pages 1–6, 2020.

89 Z. Javed, K. A. Yau, H. Mohamad, N. Ramli, J. Qadir, and Q. Ni. RL-Budget: A learning-based cluster size adjustment scheme for cognitive radio networks. *IEEE Access*, 6:1055–1072, 2018.

90 J. Jia and N. Z. Gong. Attriguard: A practical defense against attribute inference attacks via adversarial machine learning. In *27th USENIX Security Symposium (USENIX Security 18)*, pages 513–529, 2018.

91 Z. Jin, K. Yao, B. Lee, J. Cho, and L. Zhang. Channel status learning for cooperative spectrum sensing in energy-restricted cognitive radio networks. *IEEE Access*, 7:64946–64954, 2019.

92 Y. Jong. An efficient global optimization algorithm for nonlinear sum-of-ratios problem. *Optimization Online*, pages 1–21, 2012.

93 C. Karakus and S. Diggavi. Enhancing multiuser MIMO through opportunistic D2D cooperation. *IEEE Transactions on Wireless Communications*, 16(9):5616–5629, 2017.

94 A. Kaur and K. Kumar. Energy-efficient resource allocation in cognitive radio networks under cooperative multi-agent model-free reinforcement learning schemes. *IEEE Transactions on Network and Service Management*, 17(3):1337–1348, 2020.

95 A. Kaur and K. Kumar. Imperfect CSI based intelligent dynamic spectrum management using cooperative reinforcement learning framework in cognitive radio networks. *IEEE Transactions on Mobile Computing*, 21(5):1672–1683, 2020.

96 J.-B. Kim and I.-H. Lee. Non-orthogonal multiple access in coordinated direct and relay transmission. *IEEE Communications Letters*, 19(11):2037–2040, 2015. doi: 10.1109/LCOMM.2015.2474856.

97 B. Kimy, S. Lim, H. Kim, S. Suh, J. Kwun, S. Choi, C. Lee, S. Lee, and D. Hong. Non-orthogonal multiple access in a downlink multiuser beamforming system. In *MILCOM 2013 - 2013 IEEE Military Communications Conference*, pages 1278–1283, 2013. doi: 10.1109/MILCOM.2013.218.

98 T. A. Le, Q.-T. Vien, H. X. Nguyen, D. W. K. Ng, and R. Schober. Robust chance-constrained optimization for power-efficient and secure SWIPT systems. *IEEE Transactions on Green Communications and Networking*, 1(3):333–346, 2017. doi: 10.1109/TGCN.2017.2706063.

99 Y. Li and Q. Peng. Achieving secure spectrum sensing in presence of malicious attacks utilizing unsupervised machine learning. In *MILCOM 2016 - 2016 IEEE Military Communications Conference*, pages 174–179, 2016.

100 J. C. F. Li, M. Lei, and F. Gao. Device-to-device (D2D) communication in MU-MIMO cellular networks. In *2012 IEEE Global Communications Conference (GLOBECOM)*, pages 3583–3587. IEEE, 2012.

101 X. Li, J. Fang, W. Cheng, H. Duan, Z. Chen, and H. Li. Intelligent power control for spectrum sharing in cognitive radios: A deep reinforcement learning approach. *IEEE Access*, 6:25463–25473, 2018.

102 T. Li, A. K. Sahu, A. Talwalkar, and V. Smith. Federated learning: Challenges, methods, and future directions. *IEEE Signal Processing Magazine*, 37(3):50–60, 2020. doi: 10.1109/MSP.2020.2975749.

103 T. Li, A. K. Sahu, M. Zaheer, M. Sanjabi, A. Talwalkar, and V. Smith. Federated optimization in heterogeneous networks. *Proceedings of Machine Learning and Systems*, 2:429–450, 2020.

104 M. Lichtman, J. D. Poston, S. Amuru, C. Shahriar, T. C. Clancy, R. M. Buehrer, and J. H. Reed. A communications jamming taxonomy. *IEEE Security Privacy*, 14(1):47–54, 2016.

105 C. Lim, T. Yoo, B. Clerckx, B. Lee, and B. Shim. Recent trend of multiuser MIMO in LTE-advanced. *IEEE Communications Magazine*, 51(3):127–135, 2013.

106 Y. Lin, S. Han, H. Mao, Y. Wang, and W. J. Dally. Deep gradient compression: Reducing the communication bandwidth for distributed training. *arXiv preprint arXiv:1712.01887*, 2017.

107 V. Liu, A. Parks, V. Talla, S. Gollakota, D. Wetherall, and J. R. Smith. Ambient backscatter: Wireless communication out of thin air. In *Proceedings of the ACM SIGCOMM 2013 Conference on SIGCOMM*, SIGCOMM '13, pages 39–50, New York, NY, USA, 2013. Association for Computing Machinery. ISBN 9781450320566.

108 Y. Liu, Z. Ding, M. Eïkashlan, and H. V. Poor. Cooperative non-orthogonal multiple access in 5G systems with SWIPT. In *2015 23rd European Signal Processing Conference (EUSIPCO)*, pages 1999–2003. IEEE, 2015.

109 J. Liu, H. Nishiyama, N. Kato, and J. Guo. On the outage probability of device-to-device-communication-enabled multichannel cellular networks: An RSS-threshold-based perspective. *IEEE Journal on Selected Areas in Communications*, 34(1):163–175, 2016. doi: 10.1109/JSAC.2015.2452492.

110 Y. Liu, Z. Ding, M. Elkashlan, and H. V. Poor. Cooperative non-orthogonal multiple access with simultaneous wireless information and power transfer. *IEEE Journal on Selected Areas in Communications*, 34(4):938–953, 2016. doi: 10.1109/JSAC.2016.2549378.

111 Y. Liu, Z. Ding, M. Elkashlan, and J. Yuan. Nonorthogonal multiple access in large-scale underlay cognitive radio networks. *IEEE Transactions on Vehicular Technology*, 65(12):10152–10157, 2016. doi: 10.1109/TVT.2016.2524694.

112 J. Liu, Y. Shi, Z. M. Fadlullah, and N. Kato. Space-air-ground integrated network: A survey. *IEEE Communications Surveys & Tutorials*, 20(4):2714–2741, 2018. doi: 10.1109/COMST.2018.2841996.

113 M. Liu, T. Song, L. Zhang, H. Sari, and G. Gui. Multi-efficiency based resource allocation for cognitive radio networks with deep learning. In *2018 IEEE 10th Sensor Array and Multichannel Signal Processing Workshop (SAM)*, pages 56–59, 2018.

114 X. Liu, Y. Xu, L. Jia, Q. Wu, and A. Anpalagan. Anti-jamming communications using spectrum waterfall: A deep reinforcement learning approach. *IEEE Communications Letters*, 22(5):998–1001, 2018.

115 M. Liu, T. Song, J. Hu, J. Yang, and G. Gui. Deep learning-inspired message passing algorithm for efficient resource allocation in cognitive radio networks. *IEEE Transactions on Vehicular Technology*, 68(1):641–653, 2019.

116 S. Liu, Y. Xu, X. Chen, X. Wang, M. Wang, W. Li, Y. Li, and Y. Xu. Pattern-aware intelligent anti-jamming communication: A sequential deep reinforcement learning approach. *IEEE Access*, 7:169204–169216, 2019.

117 F. Liu, Y. Cui, C. Masouros, J. Xu, T. X. Han, Y. C. Eldar, and S. Buzzi. Integrated sensing and communications: Toward dual-functional wireless networks for 6G and beyond. *IEEE Journal on Selected Areas in Communications*, 40(6):1728–1767, 2022.

118 X. Lu, P. Wang, D. Niyato, D. I. Kim, and Z. Han. Wireless networks with RF energy harvesting: A contemporary survey. *IEEE Communications Surveys & Tutorials*, 17(2):757–789, 2014.

119 L. Lv, Q. Ni, Z. Ding, and J. Chen. Application of non-orthogonal multiple access in cooperative spectrum-sharing networks over nakagami-m fading channels. *IEEE Transactions on Vehicular Technology*, 66(6):5506–5511, 2017. doi: 10.1109/TVT.2016.2627559.

120 S. Ma, Y. Zhu, G. Wang, and R. He. Machine learning aided channel estimation for ambient backscatter communication systems. In *2018 IEEE International Conference on Communication Systems (ICCS)*, pages 67–71, 2018.

121 C. Ma, J. Li, M. Ding, H. H. Yang, F. Shu, T. Q. S. Quek, and H. V. Poor. On safeguarding privacy and security in the framework of federated learning. *IEEE Network*, 34(4):242–248, 2020. doi: 10.1109/MNET.001.1900506.

122 X. Ma, H. Sun, and R. Q. Hu. Scheduling policy and power allocation for federated learning in NOMA based MEC. In *GLOBECOM 2020 - 2020 IEEE Global Communications Conference*, Taipei, Taiwan, pages 1–7, 2020. doi: 10.1109/GLOBECOM42002.2020.9322270.

123 S. Machuzak and S. K. Jayaweera. Reinforcement learning based anti-jamming with wideband autonomous cognitive radios. In *2016 IEEE/CIC International Conference on Communications in China (ICCC)*, pages 1–5, 2016.

124 V. Maglogiannis, D. Naudts, A. Shahid, S. Giannoulis, E. Laermans, and I. Moerman. Cooperation techniques between LTE in unlicensed spectrum and Wi-Fi towards fair spectral efficiency. *Sensors*, 17(9):1994, 2017.

125 M. Mahmoudi, K. Faez, and A. Ghasemi. Defense against primary user emulation attackers based on adaptive Bayesian learning automata in cognitive radio networks. *Ad Hoc Networks*, 102:102147, 2020. ISSN 1570-8705.

126 Y. Mao, J. Zhang, and K. B. Letaief. Dynamic computation offloading for mobile-edge computing with energy harvesting devices. *IEEE Journal on Selected Areas in Communications*, 34(12):3590–3605, 2016. doi: 10.1109/JSAC.2016 .2611964.

127 M. Massaro and F. Beltrán. Will 5G lead to more spectrum sharing? Discussing recent developments of the LSA and the CBRS spectrum sharing frameworks. *Telecommunications Policy*, 44(7):101973, 2020. ISSN 0308-5961.

128 B. McMahan, E. Moore, D. Ramage, S. Hampson, and B. A. y Arcas. Communication-efficient learning of deep networks from decentralized data. In *Artificial Intelligence and Statistics*, pages 1273–1282. PMLR, 2017.

129 J. Men and J. Ge. Non-orthogonal multiple access for multiple-antenna relaying networks. *IEEE Communications Letters*, 19(10):1686–1689, 2015.

130 G. J. Mendis, J. Wei, and A. Madanayake. Deep learning-based automated modulation classification for cognitive radio. In *2016 IEEE International Conference on Communication Systems (ICCS)*, pages 1–6, 2016.

131 M. Zhou, J. Shen, H. Chen, and L. Xie. A cooperative spectrum sensing scheme based on the Bayesian reputation model in cognitive radio networks. In *2013 IEEE Wireless Communications and Networking Conference (WCNC)*, pages 614–619, 2013.

132 J. Mitola and G. Q. Maguire. Cognitive radio: Making software radios more personal. *IEEE Personal Communications*, 6(4):13–18, 1999.

133 M. Mizmizi, M. Brambilla, D. Tagliaferri, C. Mazzucco, M. Debbah, T. Mach, R. Simeone, S. Mandelli, V. Frascolla, R. Lombardi, et al. 6G V2X technologies and orchestrated sensing for autonomous driving. *arXiv preprint arXiv:2106.16146*, 2021.

134 P. Mohassel and Y. Zhang. SecureML: A system for scalable privacy-preserving machine learning. In *2017 IEEE Symposium on Security and Privacy (SP)*, pages 19–38, 2017. doi: 10.1109/SP.2017.12.

135 L. Mohjazi, I. Ahmed, S. Muhaidat, M. Dianati, and M. Al-Qutayri. Downlink beamforming for SWIPT multi-user MISO underlay cognitive radio networks. *IEEE Communications Letters*, 21(2):434–437, 2016.

136 M. Mueck, M. Dolores (Lola) Pérez Guirao, R. Yallapragada, and S. Srikanteswara. Regulation and standardization activities related to spectrum sharing. *Spectrum Sharing: The Next Frontier in Wireless Networks*, pages 17–33, 2020.

137 M. Nasr, R. Shokri, and A. Houmansadr. Machine learning with membership privacy using adversarial regularization. In *Proceedings of the 2018 ACM SIGSAC Conference on Computer and Communications Security*, pages 634–646, 2018.

138 D. W. K. Ng, E. S. Lo, and R. Schober. Multiobjective resource allocation for secure communication in cognitive radio networks with wireless information and power transfer. *IEEE Transactions on Vehicular Technology*, 65(5):3166–3184, 2015.

139 G. Nie, G. Ding, L. Zhang, and Q. Wu. Byzantine defense in collaborative spectrum sensing via Bayesian learning. *IEEE Access*, 5:20089–20098, 2017.

140 N. Nonaka, A. Benjebbour, and K. Higuchi. System-level throughput of NOMA using intra-beam superposition coding and SIC in MIMO downlink when channel estimation error exists. In *2014 IEEE International Conference on Communication Systems*, pages 202–206, 2014. doi: 10.1109/ICCS.2014.7024794.

141 N. Nonaka, Y. Kishiyama, and K. Higuchi. Non-orthogonal multiple access using intra-beam superposition coding and SIC in base station cooperative MIMO cellular downlink. In *2014 IEEE 80th Vehicular Technology Conference (VTC2014-Fall)*, pages 1–5, 2014. doi: 10.1109/VTCFall.2014.6966073.

142 M. Othman, S. A. Madani, S. U. Khan, and A. ur Rehman Khan. A survey of mobile cloud computing application models. *IEEE Communications Surveys & Tutorials*, 16(1):393–413, 2013.

143 G. Pan, H. Lei, Y. Deng, L. Fan, J. Yang, Y. Chen, and Z. Ding. On secrecy performance of MISO Swipt systems with TAS and imperfect CSI. *IEEE Transactions on Communications*, 64(9):3831–3843, 2016. doi: 10.1109/TCOMM.2016.2573822.

144 C. B. Papadias, T. Ratnarajah, and D. T. M. Slock. Introduction: From cognitive radio to modern spectrum sharing. *Spectrum Sharing: The Next Frontier in Wireless Networks*, pages 1–15, 2020.

145 J. Park, J. H. Reed, A. A. Beex, T. C. Clancy, V. Kumar, and B. Bahrak. Security and enforcement in spectrum sharing. *Proceedings of the IEEE*, 102(3):270–281, 2014.

146 I. Parvez, M. G. S. Sriyananda, İ. Güvenç, M. Bennis, and A. Sarwat. CBRS spectrum sharing between LTE-U and WiFi: A multiarmed bandit approach. *Mobile Information Systems*, 2016:5909801, 2016.

147 K. Pathak and A. Banerjee. Harvest-or-transmit policy for cognitive radio networks: A learning theoretic approach. *IEEE Transactions on Green Communications and Networking*, 3(4):997–1011, 2019.

148 Q. Peng, A. Gilman, N. Vasconcelos, P. C. Cosman, and L. B. Milstein. Robust deep sensing through transfer learning in cognitive radio. *IEEE Wireless Communications Letters*, 9(1):38–41, 2020.

149 L. P. Qian, Y. J. Zhang, and J. Huang. MAPEL: Achieving global optimality for a non-convex wireless power control problem. *IEEE Transactions on Wireless Communications*, 8(3):1553–1563, 2009. doi: 10.1109/TWC.2009.080649.

150 Z. Qin, X. Tao, J. Lu, W. Tong, and G. Y. Li. Semantic communications: Principles and challenges, 2022. URL https://arxiv.org/abs/2201.01389.

151 A. Rahmati and H. Dai. Reinforcement learning for interference avoidance game in RF-powered backscatter communications. In *ICC 2019 - 2019 IEEE International Conference on Communications (ICC)*, pages 1–6, 2019.

152 V. Raj, I. Dias, T. Tholeti, and S. Kalyani. Spectrum access in cognitive radio using a two-stage reinforcement learning approach. *IEEE Journal of Selected Topics in Signal Processing*, 12(1):20–34, 2018.

153 T. G. Rodrigues, K. Suto, H. Nishiyama, and N. Kato. Hybrid method for minimizing service delay in edge cloud computing through VM migration and transmission power control. *IEEE Transactions on Computers*, 66(5):810–819, 2017. doi: 10.1109/TC.2016.2620469.

154 T. G. Rodrigues, K. Suto, H. Nishiyama, N. Kato, and K. Temma. Cloudlets activation scheme for scalable mobile edge computing with transmission power control and virtual machine migration. *IEEE Transactions on Computers*, 67(9):1287–1300, 2018. doi: 10.1109/TC.2018.2818144.

155 M. Romanelli, K. Chatzikokolakis, and C. Palamidessi. Optimal obfuscation mechanisms via machine learning. In *2020 IEEE 33rd Computer Security Foundations Symposium (CSF)*, pages 153–168, 2020.

156 D. Roy, T. Mukherjee, M. Chatterjee, and E. Pasiliao. Defense against PUE attacks in DSA networks using GAN based learning. In *2019 IEEE Global Communications Conference (GLOBECOM)*, pages 1–6, 2019.

157 Y. Sagduyu, Y. Shi, and T. Erpek. Adversarial deep learning for over-the-air spectrum poisoning attacks. *IEEE Transactions on Mobile Computing*, 20(2):306–319, 2019.

158 Y. Saito, A. Benjebbour, Y. Kishiyama, and T. Nakamura. System-level performance evaluation of downlink non-orthogonal multiple access (NOMA). In *2013 IEEE 24th Annual International Symposium on Personal, Indoor, and Mobile Radio Communications (PIMRC)*, pages 611–615, 2013. doi: 10.1109/PIMRC.2013 .6666209.

159 Y. Saito, Y. Kishiyama, A. Benjebbour, T. Nakamura, A. Li, and K. Higuchi. Non-orthogonal multiple access (NOMA) for cellular future radio access. In *2013 IEEE 77th Vehicular Technology Conference (VTC Spring)*, pages 1–5, 2013. doi: 10 .1109/VTCSpring.2013.6692652.

160 R. Sarmah, A. Taggu, and N. Marchang. Detecting Byzantine attack in cognitive radio networks using machine learning. *Wireless Networks*, 26:5939–5950, 2020. ISSN 1572-8196.

161 F. Shah-Mohammadi and A. Kwasinski. Deep reinforcement learning approach to QoE-driven resource allocation for spectrum underlay in cognitive radio networks. In *2018 IEEE International Conference on Communications Workshops (ICC Workshops)*, pages 1–6, 2018.

162 Y. Sharaf-Dabbagh and W. Saad. Transfer learning for device fingerprinting with application to cognitive radio networks. In *2015 IEEE 26th Annual International Symposium on Personal, Indoor, and Mobile Radio Communications (PIMRC)*, pages 2138–2142, 2015.

163 Y. Shi, Y. E. Sagduyu, T. Erpek, K. Davaslioglu, Z. Lu, and J. H. Li. Adversarial deep learning for cognitive radio security: Jamming attack and defense strategies. In *2018 IEEE International Conference on Communications Workshops (ICC Workshops)*, pages 1–6, 2018.

164 S. Singh and A. Trivedi. Anti-jamming in cognitive radio networks using reinforcement learning algorithms. In *2012 Ninth International Conference on Wireless and Optical Communications Networks (WOCN)*, pages 1–5, 2012.

165 F. Slimeni, B. Scheers, Z. Chtourou, and V. Le Nir. Jamming mitigation in cognitive radio networks using a modified Q-learning algorithm. In *2015 International Conference on Military Communications and Information Systems (ICMCIS)*, pages 1–7, 2015.

166 F. Slimeni, B. Scheers, Z. Chtourou, V. Le Nir, and R. Attia. Cognitive radio jamming mitigation using Markov decision process and reinforcement learning. *Procedia Computer Science*, 73:199–208, 2015.

167 Q. H. Spencer, A. L. Swindlehurst, and M. Haardt. Zero-forcing methods for downlink spatial multiplexing in multiuser MIMO channels. *IEEE Transactions on Signal Processing*, 52(2):461–471, 2004.

168 S. Srinivasan, K. B. Shivakumar, and M. Mohammad. Semi-supervised machine learning for primary user emulation attack detection and prevention through core-based analytics for cognitive radio networks. *International Journal of Distributed Sensor Networks*, 15(9), 2019. doi: 10.1177/1550147719860365.

169 P. Subbulakshmi and M. Prakash. Mitigating eavesdropping by using fuzzy based MDPOP-Q learning approach and multilevel Stackelberg game theoretic approach in wireless CRN. *Cognitive Systems Research*, 52:853–861, 2018.

170 Q. Sun, S. Han, I. Chin-Lin, and Z. Pan. On the ergodic capacity of MIMO NOMA systems. *IEEE Wireless Communications Letters*, 4(4):405–408, 2015. doi: 10.1109/LWC.2015.2426709.

171 Q. Sun, S. Han, Z. Xu, S. Wang, I. Chih-Lin, and Z. Pan. Sum rate optimization for MIMO non-orthogonal multiple access systems. In *2015 IEEE Wireless Communications and Networking Conference (WCNC)*, pages 747–752. IEEE, 2015.

172 H. Sun, B. Xie, R. Q. Hu, and G. Wu. Non-orthogonal multiple access with SIC error propagation in downlink wireless MIMO networks. In *2016 IEEE 84th Vehicular Technology Conference (VTC-Fall)*, pages 1–5. IEEE, 2016.

173 H. Sun, F. Zhou, and R. Q. Hu. Joint offloading and computation energy efficiency maximization in a mobile edge computing system. *IEEE Transactions on Vehicular Technology*, 68(3):3052–3056, 2019. doi: 10.1109/TVT.2019.2893094.

174 H. Sun, F. Zhou, R. Q. Hu, and L. Hanzo. Robust beamforming design in a NOMA cognitive radio network relying on SWIPT. *IEEE Journal on Selected Areas in Communications*, 37(1):142–155, 2018. doi: 10.1109/JSAC.2018.2872375.

175 H. Sun, Q. Wang, R. Q. Hu, and Y. Qian. Outage probability study in a NOMA relay system. In *2017 IEEE Wireless Communications and Networking Conference (WCNC)*, San Francisco, CA, USA, pages 1–6, 2017. doi: 10.1109/WCNC.2017.7925775.

176 H. Sun, Q. Wang, S. Ahmed, and R. Q. Hu. Non-orthogonal multiple access in a mmWave based IoT wireless system with SWIPT. In *2017 IEEE 85th Vehicular Technology Conference (VTC Spring)*, Sydney, NSW, Australia, pages 1–5, 2017. doi: 10.1109/VTCSpring.2017.8108186.

177 H. Sun, Q. Wang, X. Ma, Y. Xu, and R. Q. Hu. Towards green mobile edge computing offloading systems with security enhancement. In *2020 Intermountain Engineering, Technology and Computing (IETC)*, Orem, UT, USA, pages 1–6, 2020. doi: 10.1109/IETC47856.2020.9249092.

178 H. Sun, X. Ma, and R. Q. Hu. Adaptive federated learning with gradient compression in uplink NOMA. *IEEE Transactions on Vehicular Technology*, 69(12):16325–16329, 2020. doi: 10.1109/TVT.2020.3027306.

179 H. Sun, Y. Xu, and R. Q. Hu. A NOMA and MU-MIMO supported cellular network with underlaid D2D communications. In *2016 IEEE 83rd Vehicular Technology Conference (VTC Spring)*, Nanjing, China, pages 1–5, 2016. doi: 10.1109/VTCSpring.2016.7504086.

180 J. Tan, L. Zhang, Y. Liang, and D. Niyato. Intelligent sharing for LTE and WiFi systems in unlicensed bands: A deep reinforcement learning approach. *IEEE Transactions on Communications*, 68(5):2793–2808, 2020.

181 F. Tang, Z. M. Fadlullah, N. Kato, F. Ono, and R. Miura. AC-POCA: Anticoordination game based partially overlapping channels assignment in combined UAV and D2D-based networks. *IEEE Transactions on Vehicular Technology*, 67(2):1672–1683, 2018. doi: 10.1109/TVT.2017.2753280.

182 C. Tarver, M. Tonnemacher, V. Chandrasekhar, H. Chen, B. L. Ng, J. Zhang, J. R. Cavallaro, and J. Camp. Enabling a "use-or-share" framework for PAL-GAA sharing in CBRS networks via reinforcement learning. *IEEE Transactions on Cognitive Communications and Networking*, 5(3):716–729, 2019.

183 R. H. Tehrani, S. Vahid, D. Triantafyllopoulou, H. Lee, and K. Moessner. Licensed spectrum sharing schemes for mobile operators: A survey and outlook. *IEEE Communications Surveys & Tutorials*, 18(4):2591–2623, 2016.

184 P. D. Thanh, T. N. K. Hoan, H. Vu-Van, and I. Koo. Efficient attack strategy for legitimate energy-powered eavesdropping in tactical cognitive radio networks. *Wireless Networks*, 25(6):3605–3622, 2019.

185 M. Troglia, J. Melcher, Y. Zheng, D. Anthony, A. Yang, and T. Yang. FaIR: Federated incumbent detection in CBRS band. In *2019 IEEE International Symposium on Dynamic Spectrum Access Networks (DySPAN)*, pages 1–6, 2019.

186 P. V. Tuan and I. Koo. Optimal multiuser MISO beamforming for power-splitting SWIPT cognitive radio networks. *IEEE Access*, 5:14141–14153, 2017.

187 P. R. Vaka. Security and Performance Issues in Spectrum Sharing between Disparate Wireless Networks. PhD thesis, Virginia Tech, 2017.

188 N. Van Huynh, D. T. Hoang, X. Lu, D. Niyato, P. Wang, and D. I. Kim. Ambient backscatter communications: A contemporary survey. *IEEE Communications Surveys & Tutorials*, 20(4):2889–2922, 2018.

189 N. Van Huynh, D. T. Hoang, D. N. Nguyen, E. Dutkiewicz, D. Niyato, and P. Wang. Optimal and low-complexity dynamic spectrum access for RF-powered ambient backscatter system with online reinforcement learning. *IEEE Transactions on Communications*, 67(8):5736–5752, 2019.

190 N. Van Huynh, D. N. Nguyen, D. T. Hoang, and E. Dutkiewicz. Jam me if you can: Defeating jammer with deep dueling neural network architecture and ambient backscattering augmented communications. *IEEE Journal on Selected Areas in Communications*, 37(11):2603–2620, 2019.

191 N. Van Huynh, D. N. Nguyen, D. Thai Hoang, E. Dutkiewicz, and M. Mueck. Ambient backscatter: A novel method to defend jamming attacks for wireless networks. *IEEE Wireless Communications Letters*, 9(2):175–178, 2020.

192 S. Vimal, M. Khari, R. G. Crespo, L. Kalaivani, N. Dey, and M. Kaliappan. Energy enhancement using multiobjective ant colony optimization with double Q learning algorithm for IoT based cognitive radio networks. *Computer Communications*, 154:481–490, 2020.

193 B. Wang, Y. Wu, K. J. R. Liu, and T. C. Clancy. An anti-jamming stochastic game for cognitive radio networks. *IEEE Journal on Selected Areas in Communications*, 29(4):877–889, 2011. doi: 10.1109/JSAC.2011.110418.

194 K.-Y. Wang, A. M.-C. So, T.-H. Chang, W.-K. Ma, and C.-Y. Chi. Outage constrained robust transmit optimization for multiuser MISO downlinks: Tractable approximations by conic optimization. *IEEE Transactions on Signal Processing*, 62(21):5690–5705, 2014. doi: 10.1109/TSP.2014.2354312.

195 Q. Wang, H. Sun, R. Q. Hu, and A. Bhuyan. When machine learning meets spectrum sharing security: methodologies and challenges. *IEEE Open Journal of the Communications Society*, 3:176–208, 2022. doi: 10.1109/OJCOMS.2022.3146364.

196 Q. Wang, K. Ren, P. Ning, and S. Hu. Jamming-resistant multiradio multichannel opportunistic spectrum access in cognitive radio networks. *IEEE Transactions on Vehicular Technology*, 65(10):8331–8344, 2016.

197 Y. Wang, M. Sheng, X. Wang, L. Wang, and J. Li. Mobile-edge computing: Partial computation offloading using dynamic voltage scaling. *IEEE Transactions on Communications*, 64(10):4268–4282, 2016. doi: 10.1109/TCOMM.2016.2599530.

198 Y. Wang, Y. Wang, F. Zhou, Y. Wu, and H. Zhou. Resource allocation in wireless powered cognitive radio networks based on a practical non-linear energy harvesting model. *IEEE Access*, 5:17618–17626, 2017. doi: 10.1109/ACCESS.2017.2719704.

199 F. Wang, J. Xu, X. Wang, and S. Cui. Joint offloading and computing optimization in wireless powered mobile-edge computing systems. *IEEE Transactions on Wireless Communications*, 17(3):1784–1797, 2018. doi: 10.1109/TWC.2017.2785305.

200 X. Wang, R. Duan, H. Yigitler, E. Menta, and R. Jantti. Machine learning-assisted detection for BPSK-modulated ambient backscatter communication systems. In *2019 IEEE Global Communications Conference (GLOBECOM)*, pages 1–6, 2019.

201 Z. Wang, M. Song, Z. Zhang, Y. Song, Q. Wang, and H. Qi. Beyond inferring class representatives: User-level privacy leakage from federated learning. In *IEEE INFOCOM 2019 - IEEE Conference on Computer Communications*, pages 2512–2520, 2019. doi: 10.1109/INFOCOM.2019.8737416.

202 N. Wang, J. Le, W. Li, L. Jiao, Z. Li, and K. Zeng. Privacy protection and efficient incumbent detection in spectrum sharing based on federated learning. In *2020 IEEE Conference on Communications and Network Security (CNS)*, pages 1–9, 2020.

203 Q. Wang, F. Zhou, R. Q. Hu, and Y. Qian. Energy efficient robust beamforming and cooperative jamming design for IRS-assisted MISO networks. *IEEE Transactions on Wireless Communications*, 20(4):2592–2607, 2021. doi: 10.1109/TWC.2020.3043325.

204 L. Wei, R. Q. Hu, T. He, and Y. Qian. Device-to-device (D2D) communications underlaying MU-MIMO cellular networks. In *2013 IEEE Global Communications Conference (GLOBECOM)*, pages 4902–4907. IEEE, 2013.

205 X. Wen, S. Bi, X. Lin, L. Yuan, and J. Wang. Throughput maximization for ambient backscatter communication: A reinforcement learning approach. In *2019 IEEE 3rd Information Technology, Networking, Electronic and Automation Control Conference (ITNEC)*, pages 997–1003, 2019.

206 Y. Wu, B. Wang, K. J. R. Liu, and T. C. Clancy. Anti-jamming games in multi-channel cognitive radio networks. *IEEE Journal on Selected Areas in Communications*, 30(1):4–15, 2012.

207 Y. Wu, F. Hu, S. Kumar, Y. Zhu, A. Talari, N. Rahnavard, and J. D. Matyjas. A learning-based QoE-driven spectrum handoff scheme for multimedia transmissions over cognitive radio networks. *IEEE Journal on Selected Areas in Communications*, 32(11):2134–2148, 2014.

208 Y. Wu, R. Schober, D. W. K. Ng, C. Xiao, and G. Caire. Secure massive MIMO transmission in the presence of an active eavesdropper. In *2015 IEEE International Conference on Communications (ICC)*, pages 1434–1440, 2015. doi: 10.1109/ICC.2015.7248525.

209 Q. Wu, W. Chen, D. W. K. Ng, J. Li, and R. Schober. User-centric energy efficiency maximization for wireless powered communications. *IEEE Transactions on Wireless Communications*, 15(10):6898–6912, 2016. doi: 10.1109/TWC.2016.2593440.

210 Y. Wu, F. Hu, Y. Zhu, and S. Kumar. Optimal spectrum handoff control for CRN based on hybrid priority queuing and multi-teacher apprentice learning. *IEEE Transactions on Vehicular Technology*, 66(3):2630–2642, 2017.

211 Y. Wu, Y. Wang, F. Zhou, and R. Q. Hu. Computation efficiency maximization in OFDMA-based mobile edge computing networks. *IEEE Communications Letters*, 24(1):159–163, 2020. doi: 10.1109/LCOMM.2019.2950013.

212 K. Xiong, B. Wang, and K. J. R. Liu. Rate-energy region of SWIPT for MIMO broadcasting under nonlinear energy harvesting model. *IEEE Transactions on Wireless Communications*, 16(8):5147–5161, 2017. doi: 10.1109/TWC.2017.2706277.

213 C. Xu, Q. Zhang, Q. Li, Y. Tan, and J. Qin. Robust transceiver design for wireless information and power transmission in underlay MIMO cognitive radio networks. *IEEE Communications Letters*, 18(9):1665–1668, 2014. doi: 10.1109/LCOMM.2014 .2340851.

214 P. Xu, Z. Ding, X. Dai, and H. V. Poor. NOMA: An information theoretic perspective. *arXiv preprint arXiv:1504.07751*, 2015.

215 Y. Xu, H. Sun, R. Q. Hu, and Y. Qian. Cooperative non-orthogonal multiple access in heterogeneous networks. In *2015 IEEE Global Communications Conference (GLOBECOM)*, pages 1–6. IEEE, 2015.

216 Y. Xu, C. Shen, Z. Ding, X. Sun, S. Yan, G. Zhu, and Z. Zhong. Joint beamforming and power-splitting control in downlink cooperative SWIPT NOMA systems. *IEEE Transactions on Signal Processing*, 65(18):4874–4886, 2017.

217 Y. Xu, F. Yin, W. Xu, C.-H. Lee, J. Lin, and S. Cui. Scalable learning paradigms for data-driven wireless communication. *IEEE Communications Magazine*, 58(10):81–87, 2020.

218 A. Yamada, T. Nishio, M. Morikura, and K. Yamamoto. Machine learning-based primary exclusive region update for database-driven spectrum sharing. In *2017 IEEE 85th Vehicular Technology Conference (VTC Spring)*, pages 1–5, 2017.

219 Z. Yang, Z. Ding, P. Fan, and N. Al-Dhahir. The impact of power allocation on cooperative non-orthogonal multiple access networks with SWIPT. *IEEE Transactions on Wireless Communications*, 16(7):4332–4343, 2017. doi: 10.1109/TWC.2017 .2697380.

220 L. Yang, H. Zhang, M. Li, J. Guo, and H. Ji. Mobile edge computing empowered energy efficient task offloading in 5G. *IEEE Transactions on Vehicular Technology*, 67(7):6398–6409, 2018. doi: 10.1109/TVT.2018.2799620.

221 S. Yeom, I. Giacomelli, M. Fredrikson, and S. Jha. Privacy risk in machine learning: Analyzing the connection to overfitting. In *2018 IEEE 31st Computer Security Foundations Symposium (CSF)*, pages 268–282, 2018. doi: 10.1109/CSF.2018.00027.

222 F. R. Yu, H. Tang, M. Huang, Z. Li, and P. C. Mason. Defense against spectrum sensing data falsification attacks in mobile Ad Hoc networks with cognitive radios. In *MILCOM 2009 - 2009 IEEE Military Communications Conference*, pages 1–7, 2009.

223 R. Yu, Y. Zhang, Y. Liu, S. Gjessing, and M. Guizani. Securing cognitive radio networks against primary user emulation attacks. *IEEE Network*, 30(6):62–69, 2016.

224 Y. Yu, H. Li, R. Chen, Y. Zhao, H. Yang, and X. Du. Enabling secure intelligent network with cloud-assisted privacy-preserving machine learning. *IEEE Network*, 33(3):82–87, 2019. doi: 10.1109/MNET.2019.1800362.

225 Y. Yuan and Z. Ding. Outage constrained secrecy rate maximization design with SWIPT in MIMO-CR systems. *IEEE Transactions on Vehicular Technology*, 67(6):5475–5480, 2018. doi: 10.1109/TVT.2017.2717495.

226 F. Zhang and X. Zhou. Location-oriented evolutionary games for spectrum sharing. In *2014 IEEE Global Communications Conference*, pages 1047–1052, 2014.

227 L. Zhang, G. Ding, Q. Wu, Y. Zou, Z. Han, and J. Wang. Byzantine attack and defense in cognitive radio networks: A survey. *IEEE Communications Surveys & Tutorials*, 17(3):1342–1363, 2015.

228 F. Zhang, X. Zhou, and X. Cao. Location-oriented evolutionary games for price-elastic spectrum sharing. *IEEE Transactions on Communications*, 64(9):3958–3969, 2016.

229 Z. Zhang, H. Sun, and R. Q. Hu. Downlink and uplink non-orthogonal multiple access in a dense wireless network. *IEEE Journal on Selected Areas in Communications*, 35(12):2771–2784, 2017. doi: 10.1109/JSAC.2017.2724646.

230 X. Zhang, Y. Zhong, P. Liu, F. Zhou, and Y. Wang. Resource allocation for a UAV-enabled mobile-edge computing system: Computation efficiency maximization. *IEEE Access*, 7:113345–113354, 2019. doi: 10.1109/ACCESS.2019.2935217.

231 Y. Zhang, Q. Wu, and M. Shikh-Bahaei. Ensemble learning based robust cooperative sensing in full-duplex cognitive radio networks. In *2020 IEEE International Conference on Communications Workshops (ICC Workshops)*, pages 1–6, 2020.

232 Y. Zhang, Q. Wu, and M. Shikh-Bahaei. On ensemble learning-based secure fusion strategy for robust cooperative sensing in full-duplex cognitive radio networks. *IEEE Transactions on Communications*, 68(10):6086–6100, 2020.

233 F. Zhou and R. Q. Hu. Computation efficiency maximization in wireless-powered mobile edge computing networks. *IEEE Transactions on Wireless Communications*, 19(5):3170–3184, 2020. doi: 10.1109/TWC.2020.2970920.

234 F. Zhou, N. C. Beaulieu, Z. Li, J. Si, and P. Qi. Energy-efficient optimal power allocation for fading cognitive radio channels: Ergodic capacity, outage capacity, and minimum-rate capacity. *IEEE Transactions on Wireless Communications*, 15(4):2741–2755, 2016. doi: 10.1109/TWC.2015.2509069.

235 S. Zhou, Y. Wu, Z. Ni, X. Zhou, H. Wen, and Y. Zou. DoReFa-Net: Training low bitwidth convolutional neural networks with low bitwidth gradients. *arXiv preprint arXiv:1606.06160*, 2016.

236 F. Zhou, Z. Li, J. Cheng, Q. Li, and J. Si. Robust AN-aided beamforming and power splitting design for secure MISO cognitive radio with SWIPT. *IEEE Transactions on Wireless Communications*, 16(4):2450–2464, 2017.

237 P. Zhou, W. Wei, K. Bian, D. O. Wu, Y. Hu, and Q. Wang. Private and truthful aggregative game for large-scale spectrum sharing. *IEEE Journal on Selected Areas in Communications*, 35(2):463–477, 2017. doi: 10.1109/JSAC.2017.2659099.

238 F. Zhou, N. C. Beaulieu, J. Cheng, Z. Chu, and Y. Wang. Robust max-min fairness resource allocation in sensing-based wideband cognitive radio with SWIPT: Imperfect channel sensing. *IEEE Systems Journal*, 12(3):2361–2372, 2018. doi: 10.1109/JSYST.2017.2698502.

239 Y. Zhou, F. Zhou, Y. Wu, R. Q. Hu, and Y. Wang. Subcarrier assignment schemes based on Q-learning in wideband cognitive radio networks. *IEEE Transactions on Vehicular Technology*, 69(1):1168–1172, 2020.

240 H. Zhu, T. Song, J. Wu, X. Li, and J. Hu. Cooperative spectrum sensing algorithm based on support vector machine against SSDF attack. In *2018 IEEE International Conference on Communications Workshops (ICC Workshops)*, pages 1–6, 2018.

241 P. Zhu, J. Li, D. Wang, and X. You. Machine-learning-based opportunistic spectrum access in cognitive radio networks. *IEEE Wireless Communications*, 27(1):38–44, 2020.

242 Y. Zou. Physical-layer security for spectrum sharing systems. *IEEE Transactions on Wireless Communications*, 16(2):1319–1329, 2017.

Index